Printed Circuits
Design

Other McGraw-Hill Reference Books of Interest

Handbooks

AVALONE AND BAUMEISTER • *Standard Handbook for Mechanical Engineers*

BEEMAN • *Industrial Power Systems Handbook*

COOMBS • *Basic Electronic Instrument Handbook*

COOMBS • *Printed Circuits Handbook*

CROFT AND SUMMERS • *American Electricians' Handbook*

DI GIACOMO • *VLSI Handbook*

FINK AND BEATY • *Standard Handbook for Electrical Engineers*

FINK AND CHRISTIANSEN • *Electronics Engineers' Handbook*

HARPER • *Handbook of Thick Film Hybrid Microelectronics*

HICKS • *Standard Handbook of Engineering Calculations*

INGLIS • *Electronic Communications Handbook*

JOHNSON AND JASIK • *Antenna Engineering Handbook*

JURAN • *Quality Control Handbook*

KAUFMAN AND SEIDMAN • *Handbook for Electronics Engineering Technicians*

KAUFMAN AND SEIDMAN • *Handbook of Electronics Calculations*

KURTZ • *Handbook of Engineering Economics*

STOUT • *Handbook of Microprocessor Design and Applications*

STOUT AND KAUFMAN • *Handbook of Microcircuit Design and Application*

STOUT AND KAUFMAN • *Handbook of Operational Amplifier Design*

TUMA • *Engineering Mathematics Handbook*

WILLIAMS • *Designer's Handbook of Integrated Circuits*

WILLIAMS AND TAYLOR • *Electronic Filter Design Handbook*

Electronics Technology and Packaging

COOMBS • *Printed Circuits Handbook*

COOMBS • *Printed Circuits Workbook Series (5 volumes)*

EDOSOMWAN • *Improving Productivity and Quality for Electronics Assembly*

SERAPHIM, LASKY, AND LI • *Principles of Electronic Packaging*

WOODSON • *Human Factors Reference Guide for Electronics and Computer Professionals*

Dictionaries

Dictionary of Computers

Dictionary of Electrical and Electronic Engineering

Dictionary of Engineering

Dictionary of Scientific and Technical Terms

MARKUS • *Electronics Dictionary*

Printed Circuits Design

Featuring Computer-Aided Technologies

Gerald L. Ginsberg

McGraw-Hill, Inc.

New York St. Louis San Francisco Auckland Bogotá
Caracas Hamburg Lisbon London Madrid
Mexico Milan Montreal New Delhi Paris
San Juan São Paulo Singapore
Sydney Tokyo Toronto

Library of Congress Cataloging-in-Publication Data

Ginsberg, Gerald L.
　Printed circuits design: featuring computer-aided technologies/
Gerald L. Ginsberg

　　p.　cm.
　Includes index.
　ISBN 0-07-023309-8
　1. Printed circuits—Design and construction—Data processing.
2. Computer-aided design.　I. Title.
TK7868.P7G55　1990
　621.381′531—dc20　　　　　　　　　　　　90-41312
　　　　　　　　　　　　　　　　　　　　　　　　CIP

　3　4　5　6　7　8　9　0　DOC/DOC　9　5

ISBN　0-07-023309-8

*The sponsoring editor for this book was Daniel A. Gonneau, the
editing supervisor was Dennis Gleason, the designer was Naomi
Auerbach, and the production supervisor was Pamela A. Pelton.
This book was set in Century Schoolbook. It was composed by
McGraw-Hill's Professional Publishing composition unit.*

Printed and bound by R.R. Donnelley & Sons Company.

To Freda and Jonathan

Contents

x   Contents

Preface

The past decade has seen major advances in the performance, size, and cost of electronics equipment for all types of end-product applications. Underlying these improvements has been the rapid increase in sophistication of the two most custom elements of an electronics product, namely, integrated circuits and printed circuit boards. Thus, the printed circuit board can no longer serve as the passive interconnection panel that it used to be. In addition to providing for component mounting and interconnection, it has assumed a major role in determining the active functioning of electronic circuits.

In response to the increasing demands for cost-effectively maintaining integrated circuit performance, printed circuit boards have evolved into many different types of packaging and interconnecting structures. To complicate the design process, these printed circuit structures are being implemented with a wide variety of materials and combinations of materials; they can be manufactured by several different processes; and new surface mount, chip-on-board, tape-automated bonding, and multichip module assembly technologies are being used to supplement the use of conventional through-hole mounting.

Therefore this book addresses printed circuit design technology with respect to the basic considerations and tradeoffs associated with:

- Basic interconnection technology
- Interconnecting structure selection
- Rigid/flexible material selection
- End-product electrical performance
- Through-hole assembly technology
- Surface mount assembly technology
- Automatic assembly principles

- Computer-aided methodology
- Design setup
- Layout and component placement
- Conductor routing
- Design checking and postprocessing
- Documentation requirements

The ultimate success on the printed circuit board design will depend to a great extent on appropriate implementation decisions in these areas so that the end-product assembly will be both performance-effective and cost-effective.

Gerald L. Ginsberg

Basic Considerations

1.1 Introduction[1,2]

The design and manufacture of electronic equipment involves complex conditions. One must evaluate the most cost-effective method for the design of a system based on the state-of-the-art technology, not necessarily on off-the-shelf components. With the ever-increasing need for integration of various functions, the optimum design is established only through complete component evaluation, proper design, and procurement of existing and fabricated parts. Without awareness, involvement, and proper advice and without the foresight to recognize and take advantage of state-of-the-art technology with maximum integration, a project will not meet market objectives.

Every design usually goes through several phases. These phases are listed below:

- Initial conceptual phase
- Conceptual design phase
- Preliminary design
- Product design phase
- Validation phase

Executing these phases requires several design steps. The various steps of the design process are illustrated in the flowchart shown in Fig. 1.1.

Many people with different expertise get involved during the design of an electrical assembly. For example, one person may be very skilled at surface mount assembly, but have limited understanding of circuitry function. No one can be an expert in all aspects of technology.

Generally speaking, an electrical circuit designer is responsible for

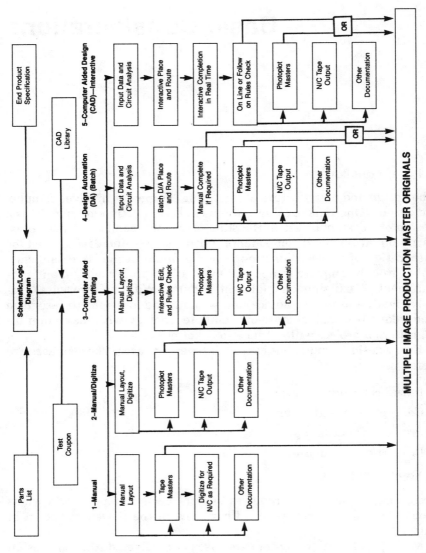

Figure 1.1 Printed circuit board design flow. *(From ANSI/IPC-D-390.[1])*

the circuit design. The circuit designer generates the initial schematic and is generally a specialist in some type of circuitry (digital logic, radiofrequency (RF) circuitry, microwave circuitry, etc.). In the final analysis, the designer is responsible for the circuitry and is the main person who evaluates the circuit performance.

The packaging designer is responsible for the physical aspects of the package and for executing the layout. The designer should review the schematic with the electrical circuit designer and assist in generating the specifications. The packaging designer should understand each specification and know how it will ultimately be measured.

In most cases, someone has fabricated a "breadboard" or "brassboard" and thus has electrically evaluated the desired circuit prior to starting the final design. In looking at the breadboard circuit for design and performance, one can begin evaluation. The first step is to consider how the circuit functions (analog, digital, power, etc.); the second step is to determine circuit density, partitioning, and the circuit relationship to the available space. Finally, the third step is to review the component specification. When the preliminary partitioning and sizing are accomplished, the remaining circuit definition items (schematic, parts list, performance requirements, etc.) can be completed.

1.2   End-Product Usage[2,3]

The success or failure of a board design depends on many interrelated considerations. From an end-product usage standpoint, as a minimum, the impact on the design on the following parameters should be considered:

- Equipment environmental conditions, such as ambient temperature, heat generated by the components, and ventilation.
- Maintenance philosophy during the service life of the equipment, especially with respect to component placement that affects component accessibility.
- Spacing between boards that might limit lead protrusions and affect the placement of brackets and hardware.
- Testing and fault location requirements that might affect component placement, conductor routing, connector contact allocations, etc.
- If an assembly is to be repairable, consideration must be given to component and circuit density and the selection of board and/or

conformal coating materials. In general, repair criteria should be in accordance with the guidelines of IPC-R-700.[4]

1.2.1 Performance, reliability, and producibility

A circuit design is assessed by considering three factors. First is the ability of the circuit to perform the functions required of it. While a great deal of evaluation can be done at the circuit level, the ability to meet system and operational specifications is the true proof of performance acceptability. Second is the ability of the circuit to perform without interruption for the operational life of the system. The circuit reliability is measured at the system level as mean time between failures (MTBF). System reliability is critical. The system should be designed to meet the end-item requirements, including life cycle, environment, and function.

Producibility, which is impossible to assess independent of the other two factors, is the third criterion. Producibility is too often given a low priority, resulting in a system which is difficult or expensive to produce. Specific manufacturing processes may complicate the problem of designing a producible system even more.

The most crucial reliability problems are plated through-hole cracking and solder joint fatigue. Both failures can be caused by the cyclic temperature strains that induce warpage of the printed circuit board or packaging as well as the interconnecting (P&I) structure and the component itself. For leaded components, the cyclic strains are accommodated to some degree by the compliancy for the component leads so that solder joint reliability is normally a less important issue.

Leadless components, on the other hand, are fully subjected to these cyclic strains, and the reliability of their solder joints needs to be carefully considered in conjunction with the other design requirements. Solder joint reliability is complicated. While the affects of some parameters are fairly well-modeled, other influences are poorly understood and so solder joint reliability remains difficult to predict.

1.2.2 Partitioning

Partitioning may be the designer's most important first step. Generally, partitioning involves two major divisions: mechanical configuration and electrical configuration.

1.2.2.1 Mechanical configurations. The first mechanical consideration is the space available for the circuit. Space is an important design consideration. The second mechanical consideration is the circuit-to-system interconnection requirements. The circuit may require fiber optics, sockets for various types of packages, RF cabling, or waveguides. Interconnection to the system is dictated by system requirements and is equal in importance to the mechanical configuration.

Third, the circuit-mounting constraints should be considered. These may include managing elevated temperatures by heat sinking, orientation of circuit, clearance between assemblies, and shielding from electromagnetic interference (EMI). Last, consider the operating environment: temperature, pressure, and stress (acceleration, shock, vibration, etc.), and any other environmental conditions.

1.2.2.2 Electrical configuration. The electrical characteristics or configuration affect partitioning. Several obvious questions pertaining to the circuitry need to be answered:

- How much of the circuitry is analog and how much is digital?
- What frequency is the circuitry (high-frequency, low-frequency, audio, or digital)?
- What are the tuning or functional trimming requirements of the circuit? This may dictate extra input/output (I/0) pins and change the partitioning scheme.
- What is the necessary power dissipation?

What current noise may be generated by parts of the circuit, and what are the effects of outside noise upon the circuit? This includes crosstalk, low-level noise, and the operating environment of the circuit.

Finally, circuit testability affects partitioning. Concern for complexity, trouble analysis, high versus low volume, and the type, amount, and complexity of the test equipment required is very important. Partitioning a circuit to provide a testable function reduces testing time, since testing one function proves the compatibility of many component tolerances.

1.2.3 Component considerations

The package dimensions, heat-transfer capability, manufacturability yield, pinout, complexity, and testability all limit the number of components that can be put into one board. Components are usually selected for electrical, thermal, and mechanical characteristics that are determined by the requirements of the package. The material composition, finish, and configuration of both the body and the terminations of a component affect the choice of assembly methods.

All components must be compatible with the assembly processes used. The physical dimensions of the component must adequately mate with the physical-handling devices of the placement equipment. The part must not be degraded, physically or electrically, by soldering or other high-temperature processes. Also, the parts must be able to tolerate the chemicals used in adhesive bonding, soldering, cleaning, or other chemical processing.

Component selection depends upon:

- Electrical characteristics
- Electrical performance
- Mating forces
- Environmental requirements
- Durability
- Repairability
- Manufacturing Methods
- Thermal Requirements
- Electrical environment
- Placement and Attachment equipment
- Polarization
- Cleaning
- Corrosion
- Part identification and Verification

1.2.4   Design standards[5,6]

Early in 1988 the Defense Electronics Supply Center (DESC) indicated to the Institute for Interconnecting and Packaging Electronic Circuits (IPC) a desire to retire the Department of Defense design standard MIL-STD-275 ("Printed Wiring for Electronic Equipment"), in favor of a standard that would be appropriate for both military and commercial applications. The document subsequently developed is identified as IPC-D-275 ("Design Standard for Rigid Printed Boards and Rigid Printed Board Assemblies") in order to maintain an easy association with its predecessor. It also supersedes the IPC's commercial printed board design documents ANSI/IPC-D-319 ("Design Standards for Rigid Single- and Double-Sided Printed Boards")[7] and ANSI/IPC-D-949 ("Design Standard for Rigid Multilayer Printed Boards").[3] However, IPC-D-275 represents a major departure from its predecessors. The most significant difference is in the format philosophy, which moves the printed board designer from the concept phase of a product development cycle through quality assurance concerns. The

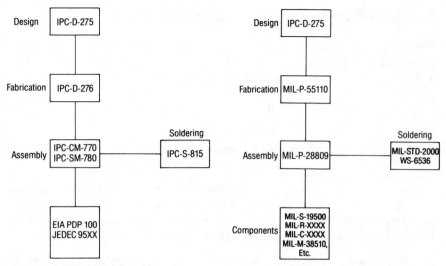

Figure 1.2 IPC rigid printed board standards and specifications. *(From IPC-D-275.[6])*

Figure 1.3 Military rigid printed board standards and specifications. *(From IPC-D-275.[6])*

designer is constantly presented with tutorial information explaining the rationale behind the stated requirement.

This philosophy serves two distinct purposes. First, it provides the designer with the ability to apply the standard to unusual design situations not specifically addressed in the document. Second, it provides an historic data trail so that future changes to the standard can be made with full knowledge as to the "why" of the requirement.

Figures 1.2 and 1.3 illustrate how IPC-D-275 fits into the commercial and military world of standards and specifications. IPC-D-275 has nine major sections that pertain to the following subjects.

1.2.4.1 Sections 1 (scope) and 2 (applicable documents). Sections 1 and 2 contain the basic "boilerplate" information that begins most standards, such as purpose, presentation, interpretation, and applicable documents. Of major significance is a system by which a printed board or printed board assembly is classified by its intended end-item use and producibility.

Producibility is related to complexity of design and the particular printed board or printed board assembly. There are three producibility levels, namely,

- General design complexity (preferred)
- Moderate design complexity (standard)

■ High-density complexity (reduced producibility)

The purpose of the producibility levels is to provide a means of communication between design and fabrication facilities. Also, end-product performance is divided into three classes, namely,

■ General electronic products
■ Dedicated-service electronic products
■ High-reliability electronic products

General electronic products. The general electronic products class includes consumer products, some computers and computer peripherals, as well as general military hardware suitable for applications where cosmetic imperfections are not important and the major requirement is the function of the completed board or board assembly.

Dedicated-service products. The dedicated-service electronic products class includes communications equipment, sophisticated business machines, instruments, and military equipment where high performance and extended life are required and for which uninterrupted service is desired but is not critical. On these products certain cosmetic imperfections are allowed.

High-reliability electronic products. The high-reliability electronic products class includes the equipment for commercial and military applications where continued performance or performance on demand is critical. Equipment downtime cannot be tolerated and the equipment must function as required, as, for example, items such as life-support or missile systems. Printed boards and printed board assemblies in this class are suitable for applications where high levels of assurance are required and service is essential. The ability to match any producibility level with any intended use classification is a unique aspect of IPC-D-275. For example, this enables the military user to take advantage of certain board constructions enjoyed by the commercial community.

1.2.4.2 Section 3 (design considerations). Section 3 of IPC-D-275 represents the first major departure for military-type standards. It provides overall considerations to be addressed at a project start meeting. Since, unfortunately, design layouts are often started with insufficient information, Section 3 outlines the various types and minimum requirements to provide a simple checklist for those persons responsible for supplying design input.

Design features. Design features, such as viewing conventions, location grid systems, and preferred dimensioning practices are discussed in detail. The intention here is to provide the basic concepts required by the designer to the appropriate personnel that are upstream and downstream of the printed board design activity. Section 3 provides expanded information concerning conductor spacing, Table 1.1.

Thermal management and mechanical requirement considerations covered include topics such as thermal matching, total board heat dissipation, support, and vibration.

Testing. It is imperative that a printed board and printed board assembly test philosophy be developed and agreed upon prior to the start of the design activity. Thus, a detailed discussion is presented that outlines the many test considerations the design review team must take into account.

Transmission lines. Section 3 also provides formulas for characteristic impedance, intrinsic line capacitance, as well as crossover capacitance for four types of transmission lines, i.e.,

- Microstrip
- Embedded microstrip
- Stripline
- Dual stripline.

Detailed tutorials enable decisions to be made with respect to the technologies that may have to be employed.

Power distribution. Considerations relating to power distribution for both digital and analog circuits are detailed, along with recommendations for circuit placement with respect to circuit speed.

Material. A material subsection provides guidance for selecting materials such as laminates, bonding agents, metallic foils and films, metallic platings and coatings, and organic protective coatings. Table 1.2, pertaining to copper foil and film requirements, illustrates one of the many tables provided in Section 3 of IPC-D-275.

1.2.4.3 Section 4 (component mounting and attachment). Section 4 describes the various types and styles of components that printed board designers might encounter and common methods for their attachment to the printed board. Thoroughly covered are

- Through-the-board mounted components
- Surface-mounted leaded components
- Surface-mounted leadless components
- Component lead sockets

TABLE 1.1 Electrical Conductor Spacing

| Voltage between conductors (dc or ac peak) | Minimum spacing, mm (in) | | | | | | |
| | Bare board | | | | | Assembly | |
	B1	B2	B3	B4	A5	A6	A7
0–15	0.1 (0.004)	0.64 (0.025)	0.64 (0.025)	0.13 (0.005)	0.13 (0.005)	0.13 (0.005)	0.13 (0.005)
16–30	0.1 (0.004)	0.64 (0.025)	0.64 (0.025)	0.13 (0.005)	0.13 (0.005)	0.25 (0.010)	0.13 (0.005)
31–50	0.1 (0.004)	0.64 (0.025)	0.64 (0.025)	0.13 (0.005)	0.13 (0.005)	0.38 (0.015)	0.13 (0.005)
51–100	0.1 (0.004)	0.64 (0.025)	1.50 (0.060)	0.13 (0.005)	0.13 (0.005)	0.5 (0.020)	0.13 (0.005)
101–150	0.2 (0.008)	0.64 (0.025)	3.18 (0.125)	0.38 (0.015)	0.038 (0.015)	0.75 (0.030)	0.38 (0.015)
151–170	0.2 (0.008)	1.27 (0.050)	3.18 (0.125)	0.38 (0.015)	0.38 (0.015)	0.75 (0.030)	0.38 (0.015)
171–250	0.2 (0.008)	1.27 (0.050)	6.4 (0.250)	0.38 (0.015)	0.38 (0.015)	0.75 (0.030)	0.38 (0.15)
251–300	0.2 (0.008)	1.27 (0.050)	12.7 (0.500)	0.38 (0.015)	0.38 (0.015)	0.75 (0.030)	0.38 (0.15)
301–500	0.3 (0.012)	2.54 (0.100)	12.7 (0.500)	0.75 (0.030)	0.75 (0.030)	1.50 (0.060)	0.75 (0.030)
Greater than 500	0.0025 (0.0001)N	0.005 (0.0002)N	0.0254 (0.001)N	0.00305 (0.00012)N	0.00305 (0.00012)N	0.00305 (0.00012)N	0.00305 (0.00012)N

NOTE: B1, internal conductors; B2, External conductors, uncoated, sea level to 10,000 ft; B3, external conductors, uncoated, over 10,000 ft; B4, external conductors, with permanent polymer coating (any elevation); A5, external conductors, with conformal coating over assembly (any elevation); A6, external component lead/termination, uncoated; A7, external component lead/termination, with conformal coating (any elevation).
SOURCE: IPC-D-275, Ref. 6.

TABLE 1.2 Copper Foil and Film Requirements

Copper type	Class 1	Class 2	Class 3
Minimum starting copper foil— external	¼ oz/ft² 9 μm (0.0004 in)	½ oz/ft² 12 μm (0.0006–0.0008 in)	½ oz/ft² 12 μm (0.0006–0.0008 in)
Minimum starting copper foil— internal	½ oz/ft² 12 μm (0.0005 in)	½ oz/ft² 12 μm (0.0005 in)	½ oz/ft² 12 μm (0.0005 in)
Starting copper film (semiadditive)	5 μm (0.0002 in)	5 μm (0.0002 in)	5 μm (0.0002 in)
Final copper film (fully additive)	15–20 μm (0.0006–0.0008 in)	15–20 μm (0.0006–0.0008 in)	15–20 μm (0.0006–0.0008 in)

*All dimensional values are nominal and derived from weight measurements.
SOURCE: IPC-D-275, Ref. 6.

- Connectors and interconnects
- Terminals
- Special wiring, such as bus bars and jumpers

Component lead stress-relief calculations, or the rationale behind them, are shown for most leaded components. An example of stress-relief requirements for planar-mounted flat packs is shown in Fig. 1.4.

Considerations for automatic component mounting are discussed in tutorial form with appropriate references to IPC-CM-770, "Guidelines for Printed Board Component Mounting."

1.2.4.4 Section 5 (board design requirements). Section 5 of IPC-D-275 is formatted in the typical board layout methodology. This section begins with printed board layout concepts and methods for making feasibility density evaluations. Board geometries, such as size, shape, dimensions, tolerances, data, dielectric spacing, and thickness considerations are discussed as they relate to final fabrication and as-

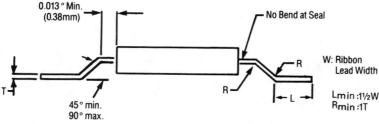

Figure 1.4 Configuration of ribbon leads for planar-mounted flat packs. *(From IPC-D-275.[6])*

TABLE 1.3 **Typical Maximum Panel Size for Manufacturing Operations**

Operation	Typical max. panel size, mm (in)
Drill	457 × 610 (18 × 24)
Scrub, deburr, and most conveyorized finishing equipment	610 × open (24 × open)
Plating equipment	Custom-sized, check with fabricator
Exposure equipment	610 × 610 (24 × 24)
Routing equipment	457 × 610 (18 × 24)
Screening equipment	508 × 762 (20 × 30)
Bare board test	457 × 457 (18 × 18)
Laminating press size (based on 610 × 760 mm (24 × 30 in) press with 51 mm (2-in) open area on platen edges)	508 × 660 (20 × 26)
Solder coating	457 × 533 (18 × 21)

SOURCE: IPC-D-275, Ref. 6.

sembly processes. Typical maximum panel sizes for various manufacturing operations are illustrated in Table 1.3.

Dimensioning. IPC-D-275 clearly defines the methods and procedures for design datum selection consistent with current industry and military dimensioning standards. The concepts are detailed in illustrations that build upon each other so as not to intimidate users that are unfamiliar with these design practices. An example of the dimensioning practices recommended in Section 5 is shown in Fig. 1.5.

All tolerance references are with respect to "true position," not bilateral (+/−). This represents a departure from what many commercial designers are accustomed to using.

Conductors and lands. A subsection on circuit features covers topics such as:

- Conductors
- Lands with holes
- Lands without holes
- Nonfunctional lands
- Large conductive areas
- Offset Lands
- Orientation symbols

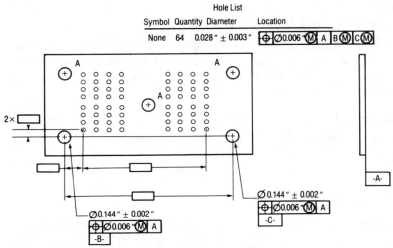

Figure 1.5 Recommended location of a pattern of plated-through holes. *(From IPC-D-275.[6])*

Highlighted in this subsection are tables containing the requirements for minimum annular ring, feature location tolerance, and minimum standard fabrication allowances for interconnection lands (Table 1.4).

Land Patterns for many surface-mounted components are defined along with the rationale behind their creation. The intent is to provide the printed board designer with the necessary tools for applying the basic principles for defining component lands that are not specifically illustrated in the standard. Other sample calculations and illustra-

TABLE 1.4 Minimum Standard Fabrication Allowance for Interconnection Lands

Greatest board, x, y dimension, mm (in)	Level A, mm (in)	Level B, mm (in)	Level C, mm (in)
Up to 300 (12.0)	0.5 (0.020)	0.4 (0.016)	0.25 (0.010)
Up to 450 (18.0)	0.6 (0.024)	0.5 (0.020)	0.45 (0.018)
Up to 600 (24.0)	0.7 (0.028)	0.6 (0.024)	0.55 (0.022)

NOTE: The above considerations shall be incorporated into the minimum land provided on the production master such that minimum land dimension = $a + 2b + c$. All lands and annular rings shall be maximized wherever feasible, consistent with good design practice and electrical clearance requirements. 1. For copper weights greater than 1 oz/ft², add 0.05 mm (0.002 in) minimum to the fabrication allowance for each additional oz/ft² of copper used. 2. For more than eight layers add 0.05 mm (0.002 in).

SOURCE: IPC-D-275, Ref. 6.

tions provided clearly define requirements for thermal relief and test point lands, to name a few.

Holes. A dedicated subsection on holes is divided into categories concerning holes that are supported (plated) and holes that are unsupported (unplated), such as printed board mounting holes. Related topics covered include:

- Plated-through hole diameter to lead diameter relationships
- Board thickness to hole diameter aspect ratios
- Blind and buried vias
- Etchback
- Hole location tolerances

These are discussed in a manner that enables the printed board designer to make decisions concerning hole-related parameters that are not specifically addressed in the standard. An example of this type of information is shown in Table 1.5.

Coatings and markings. Another subsection includes specific information on clearances for photoimageable and screenable solder masks, compatibility, and marking legend requirements.

1.2.4.5 Section 6 (documentation and phototooling). Section 6 of IPC-D-275 provides the printed board designer with the basic documentation and phototooling requirements for the

- Layout

TABLE 1.5 Minimum Hole Size for Plated-Through Holes

Board thickness, mm (in)	Class 1, mm (in)	Class 2, mm (in)	Class 3, mm (in)
< 1.0 (0.040)	C 0.15 (0.006)	C 0.2 (0.008)	C 0.25 (0.010)
1.0 → 1.6 (0.040 → 0.063)	C 0.2 (0.008)	C 0.25 (0.010)	B 0.3 (0.012)
1.6 → 2.0 (0.063 → 0.080)	C 0.3 (0.012)	B 0.4 (0.016)	B 0.5 (0.020)
> 2.0 (0.080)	B 0.4 (0.016)	A 0.5 (0.020)	A 0.6 (0.024)

NOTE: If copper in hole is greater than 0.03 mm (0.0012 in) hole size can be reduced by one class.
SOURCE: IPC-D-275, Ref. 6.

- Master drawing
- Artwork
- Production master
- Assembly drawing
- Final Schematic or logic diagram
- Bill of materials

Unlike the superseded MIL-STD-275, in Section 6 of IPC-D-275, specific references are made to other documents for documentation and artwork requirements, i.e., ANSI/IPC-D-325, "Documentation Requirements for Printed Boards," and ANSI/IPC-D-310 "Guidelines for Phototool Generation and Measurement Techniques."

1.2.4.6 Section 7 (quality assurance). The procedure for designing a printed board would not be complete without a clear understanding of quality assurance issues. Thus, Section 7 of IPC-D-275 covers such topics as

- Quality conformance evaluations
- Qualification evaluations
- Material quality assurance

The all-to-familiar evaluation coupons are illustrated and described, as well as a host of coupons designed to address both current and future technologies. Table 1.6 illustrates the variety of coupons and the frequency of their use for the three performance classes.

1.2.4.7 Section 8 (material considerations appendix). IPC-D-275 has an appendix that contains useful information with descriptions of laminate material coding in accirdance with the requirements of MIL-P-13949 ("Plastic Sheet, Laminated, Copper-Clad, for Printed Wiring").

1.3 Design Development[3,8]

There is no generally approved method for the development of a design for a printed circuit board. However, the following program sequence is useful.

1. *Schematic or Logic Diagram:*   Start with a careful design of the electrical circuit and prepare a schematic or logic diagram.
2. *Breadboard:*   Build a breadboard, and then analyze the circuit operation using worst-case component values or operating conditions.

TABLE 1.6   Coupon Frequency Recommendations

Coupon	Class 1	Class 2	Class 3
A. Plating thickness and hole solderability	Not required	Twice per panel	Twice per panel
As. Solderability (special)	Optional	Optional	Optional
B. Thermal stress (plated through-hole integrity)	Twice per panel, opposite corners	Twice per panel, opposite corners	Twice per panel, opposite corners
C. Plating and peel strength and surface solderability	Not required	Once per panel, location optional, pattern defined by artwork	Once per panel, location optional, pattern defined by artwork
Cs. Surface-mount solderability	Not required	Once per panel, location optional, pattern defined by artwork	Once per panel, location optional, pattern defined by artwork
Cb. Surface-mount bond strength	Not required	Once per panel, location optional, pattern defined by artwork	Once per panel, location optional, pattern defined by artwork
D. Electrical continuity	Not required	Once per panel, location optional, pattern defined by artwork	Once per panel, location optional, pattern defined by artwork
E. Moisture and insulation resistance	Twice per panel, opposite corners	Twice per panel, opposite corners	Twice per panel, opposite corners
F. Registration	Not required	Four per panel, opposite sides, fixed by artwork	Four per panel, opposite sides, fixed by artwork
Fe. Registration (optional)	Not required	Four per panel, opposite sides, fixed by artwork	Four per panel, opposite sides, fixed by artwork
G. Solder mask (if used)	Once per panel with solder mask, location optional	Once per panel with solder mask, location optional	Once per panel with solder mask, location optional

SOURCE: IPC-D-275, Ref. 6.

Use qualified components that have been tested for reliability. Revise the schematic as necessary.

3. *Layout:*   Give the final schematic or logic diagram and a component list to the design drafter. Select a board of a shape and size that will accommodate all components and fit the available space. Select appropriate I/0 connectors to accommodate all necessary

input/output signals; they should be compatible with space and environmental requirements.

Specific information regarding widths of conductors; spacing of conductors and lands, most suitable or critical routing of conductors, hole sizes and location, types of electrical interconnects, shape and bulk of components, distance between components, method of soldering, electroplating, and other board requirements must be supplied by the printed circuit board designer.

1.3.1 Schematic or logic diagram

The initial schematic or logic diagram designates the electrical functions and interconnectivity to be provided by the printed board and its assembly. The schematic should define, when applicable, critical circuit layout areas, shielding requirements, grounding and power distribution requirements, the allocation of test points, and any preassigned input/output connector locations. Schematic information may be generalized as hard copy or computer data (manually or automated).

1.3.2 Parts list

The initial parts list should contain a complete description of the part, quantities, manufacturer's name, reference designators, and, if required, special ordering instructions. Sometimes it is desirable to also include the cost of the items to be used for cost analysis purposes. The method of how the information is to be formatted and what must be included depends upon the standard established by each user.

It is advisable to group like items, e.g., resistors, capacitors, integrated circuits, in some sort of ascending or numerical order. This is of particular value to someone who must order the materials. The parts list must also have some sort of drawing number, depending upon the design facilities system. And, in most cases, the name of board for which it was prepared. Included also should be the names of the persons preparing, checking, and approving this document.

When nonstandard parts are used in the design, the detailed mechanical or physical description should be provided as a minimum. This information should be referenced to the schematic, so that the appropriate layout configuration can be determined.

A specification control drawing is required when nonstandard parts are used to ensure that future parts match the layout concepts.

1.3.3 Feasibility evaluation

Before the actual drawing of the layout is begun, a feasibility density evaluation should be made. This should be based on the maximum size of all parts required by the parts list, and the space they and their lands will require on the total board, exclusive of interconnection conductor routing. The total board geometry required for this mounting and termination of the components should then be compared to the total usable board area for this purpose.

Reasonable maximum values for this ratio are 70 percent for simple designs, 80 percent for moderate designs, and 90 percent for complex designs. Component density values higher than these will be a cause for concern. The lower these values are, the greater will be the chances for the completion of a successful board design.

An alternative method of feasibility density evaluation expresses board density in units of square inches per equivalent (14-pin) integrated circuit (IC). The calculation is made by dividing the total plated-through holes in the routing area by 14 to convert to equivalent 14-pin devices. This number is divided into the total routing area expressed in square inches. Reasonable maximum density values are 0.55 in^2/eq IC for Class A, 0.50 in^2/eq IC for Class B, and 0.45 in^2/eq IC for Class C. For additional density evaluations, see the Appendix.

1.3.4 Design layout[8]

The layout generation process should include a formal design review of layout details by as many affected disciplines within the company as possible, including fabrication, assembly and testing. The approval of the layout by representatives of the affected disciplines will assure that these design-related factors have been considered. The success or failure of a board design depends on many interrelated considerations. From an end-product usage standpoint, the impact on the design by the following typical parameters should be considered.

- Equipment environmental conditions, such as ambient temperature, heat generated by the components, ventilation, shock and vibration

- If an assembly is to be maintainable and repairable, consideration must be given to component and circuit density, the selection of board and conformal coating materials, and component placement for accessibility

- Installation interface which may affect size and location of mounting holes, connector locations, lead protrusion limitations, part placement, and placement of brackets and hardware

- Testing and fault location requirements that might affect component placement, conductor routing, connector contact allocations, etc.

- Process allowances, such as etch factor compensation for conductor widths, spacings, and land fabrication.
- Manufacturing limitations, such as minimum etched features, minimum plating thickness, and board shape and size
- Coating and marking requirements
- Technology, such as surface mount, through hole, multilayer, or metal core
- Board class, commercial through high-reliability
- Materials selection
- Producibility of printed circuit board assembly as it pertains to manufacturing equipment limitations.

The design layout from one board design to another should be such that designated areas are identified by function, i.e., power supply section confined to one area, analog circuits to another, logic circuits to another, etc. This will help to minimize crosstalk, simplify bare board and assembly test fixture design, and facilitate troubleshooting diagnostics. In addition, the design should

- Ensure that components have all testable points accessible from the secondary side of the board to facilitate probing with "bed-of-nails" test fixtures.
- Have feed-throughs and component holes placed away from board edges to allow adequate test fixture clearance.
- Require the board be laid out on a grid which matches the design team's testing concept.
- Allow provision for isolating parts of the circuit to facilitate testing and diagnostics.
- Where practical, group test points and jumper points in the same physical location on the board.
- Consider high-cost IC components for socketing so that parts can be easily replaced for further testing of that part.
- Surface-mounted components and their patterns require special consideration for test probe access, especially if components are mounted on both sides of the board.
- For surface-mount designs, provide optic targets (fiducials) to allow

the use of optic positioning and visual inspection equipment and methods.

1.3.5 Final design review[2]

An in-depth review should be made upon completion of the final layout. The responsible circuit and packaging designers should conduct the review with representatives from the fabrication, assembly, design, and testing departments. The following should be determined:

- Have the decisions made during the preliminary layout review been implemented?
- Does the circuit layout correspond to the circuit schematic?
- Has the circuit been organized for optimum circuit function?
- Have the guidelines for good design been followed?
- Are the package and substrate appropriate for the design and board mounting?
- Have all dimensional constraints been observed?
- Are external leads sufficient for the design and for board mounting?
- Are exit bonding lands correctly aligned with the package leads?
- Has the most appropriate film technology been selected?
- Are components correctly located and oriented?
- Can some of the components be more appropriately located?
- Have high-frequency interactions been considered? Have component and conductor locations and orientations been optimized? Has appropriate shielding been provided?
- Have cross-coupling and interference been minimized?
- Have matched resistors and components been located in close proximity to one another?
- Are lands of proper size and in the correct position?
- Have bonding areas and probe lands been provided for testing, troubleshooting, and dynamic trimming?
- Has heat spreading been adequately compensated for?
- Have high-current conductors been properly located? Are they of adequate size? Are they as short as possible?
- Are conductors routed correctly?
- Has conductor impedance been considered?
- Have crossovers been minimized?

- Has consideration been given to voltage breakdown, overstress, dc losses, surface breakdown, etc.?

- Are resistor values and sizes correct? Have nominal dimensions been used?

- Does the layout allow for easy troubleshooting?

- Has space been provided for circuit changes and growth?

- Has reliability been considered?

- Is the design economical to produce?

- Have diagonal lines been added to all resistors? Has appropriate color coding been used for all conductors?

- Have the title block, resistor tables, screen data, parts list, and notes been completed?

1.4 Packaging and Interconnecting Structures[2,8]

The basic function of printed circuitry is to provide support for circuit components and to interconnect them electrically. To achieve this, numerous packaging and interconnecting (P&I) structure types, varying in base dielectric material, conductor type, number of conductor planes, rigidity, etc., have developed. The printed circuit designer should be familiar with these variations and their effect on cost, component placement, wiring density, delivery cycles, and functional performance in order to select the P&I structure with the best combination of features for the particular requirements of the electronic combination apparatus or system being developed.

A P&I structure should be selected for optimum thermal, mechanical, and electrical system reliability. However, each candidate structure has particular advantages and disadvantages when compared to the others (see Table 1.7). Thus, no one P&IS will satisfy all of the needs of an application. Therefore, seek a compromise of properties best tailored for component attachment and circuit reliability.

P&I structures vary from basic printed circuit boards to very sophisticated supporting-core structures. However, some selection criteria are common to all structures. To aid in the selection process, Table 1.8 lists design parameters and material properties which affect system performance, regardless of the type of P&I structure. Table 1.9 lists the properties of the materials most common for these applications.

In general, a P&I structure will fit into one of four basic categories of construction: organic base material, nonorganic base material, supporting plane, and constraining core. A P&I structure should be designed in accordance with the requirements of IPC-D-249, "Flexible

TABLE 1.7 Packaging and Interconnecting Structure Comparison

Type	Major advantages	Major disadvantages	Comments
		Organic Base Substrate	
Epoxy fiberglass	Substrate size, weight, reworkable, dielectric processess, conventional board processes	Thermal conductivity x, y and z axis CTE	Because of its high x-y plane CTE, it should be limited to environments and applications with small changes in temperature and/or small packages.
Polyimide Fiberglass	Same as epoxy fiberglass plus high-temperature x-y axis CTE, substrate size, weight, reworkable, dielectric properties, high Tg	Thermal conductivity, z-axis CTE, moisture absorption	Same as epoxy fiberglass.
Epoxy Aramid Fiber	Same as epoxy fiberglass, x-y axis CTE, substrate size, lightest weight, reworkable, dielectric properties	Thermal conductivity, z-axis CTE, resin microcracking, water absorption	Volume fraction of fiber can be controlled to tailor x-y CTE. Resin selection critical to reducing resin microcracks.
Polyimide Aramid Fiber	Same as epoxy aramid fiber, x axis CTE, substrate size, weight, reworkable, dielectric properties.	Thermal conductivity, z axis CTE, resin microcracking, water absorption	Same as epoxy aramid fiber.
Polyimide Quartz (Fused Silica)	Same as polyimide aramid fiber, x-y axis CTE, substrate size, weight, reworkable, dielectric properties	Thermal conductivity, z axis CTE, drilling, availability, cost, low resin content required	Volume fraction of fiber can be controlled to tailor x-y CTE. Drill wearout higher than with fiberglass.
Fiberglass/Aramid Composite Fiber	Same as polyimide aramid fiber, no surface microcracks, z axis CTE, substrate size, weight, reworkable, dielectric properties	Thermal conductivity, x and y axis CTE, water absorption, process solution entrapment	Resin microcracks are confined to internal layers and cannot damage external circuitry.
Fiberglass/Teflon® Laminates	Dielectric constant, high temperature	Same as epoxy fiberglass, low temperature stability, thermal conductivity, x and y axis CTE	Suitable for high-speed logic applications. Same as epoxy fiberglass.

Flexible Dielectric	Light weight, minimal concern to CTE, configuration flexibility	Size, cost, z axis expansion	Rigid-flexible boards offer trade-off compromises.
Thermoplastic	3D configurations, low high-volume cost	High injection-molding setup costs.	Relatively new for these applications.
Nonorganic Base			
Alumina (Ceramic)	CTE, thermal conductivity, conventional thick-film or thin-film processing, integrated resistors	Substrate size, rework limitations, weight, cost, brittle, dielectric constant	Most widely used for hybrid circuit technology.
Supporting Plane			
Printed Board Bonded to Plane Support (Metal or Non-Metal)	Substrate size, reworkability, dielectric properties, conventional board processing, x-y axis CTE, stiffness, shielding, cooling	Weight	The thickness/CTE of the metal core can be varied along with the board thickness, to tailor the overall CTE of the composite.
Sequential Processed Board with Supporting Plane Core	Same as board bonded to supporting plane	Weight	Same as board bonded to supporting plane.
Discrete Wire	High-speed interconnections, good thermal and electrical features	Licensed process, requires special equipment	Same as board bonded to low-expansion metal support plane.
Constraining Core			
Porcelainized Copper Clad Invar	Same as alumina	Reworkability, compatible thick-film materials	Thick film materials are still under development.
Printed Board Bonded with Constraining Metal Core	Same as board bonded to low-expansion metal cores, stiffness, thermal conductivity, low weight	Cost, microcracking	The thickness of the graphite and board can be varied to tailor the overall CTE of the composite.
Compliant Layer Structures	Substrate size, dielectric properties, x-y axis, CTE	z axis CTE, thermal conductivity	Compliant layer absorbs difference in CTE between ceramic package and substrate.

source: ANSI/IPC-SM-780, Ref. 2.

TABLE 1.8 Packaging and Interconnecting Structure Considerations

Design parameters	Material properties								
	Transition temperature	Coefficient of thermal expansion	Thermal conductivity	Tensile modulus	Flexural modulus	Dielectric constant	Volume resistivity	Surface resistivity	Moisture absorption
Temperature & power cycling	X	X	X						
Vibration				X	X				
Mechanical shock				X	X				
Temperature & humidity	X	X				X	X	X	X
Power density	X		X	X					
Chip carrier size		X							
Circuit density						X	X	X	
Circuit speed						X	X	X	

SOURCE: ANSI/IPC-SM-780, Ref. 2.

TABLE 1.9 Packaging and Interconnecting Structure Material Properties

Material	Glass Transition Temperature, °C	XY coefficient of thermal expansion, PPM/°C	Thermal conductivity, W/M°C	XY tensile modulus, psi $\times 10^6$	Dielectric constant (at 1 MHz)	Volume resistivity, Ω/cm	Surface resistivity, Ω	Moisture absorption, %
Epoxy fiberglass	125	13–18	0.16	2.5	4.8	10^{12}	10^{13}	0.10
Polyimide fiberglass	250	12–16	0.35	2.8	4.8	10^{14}	10^{13}	0.35
Epoxy aramid fiber	125	6–8	0.12	4.4	3.9	10^{16}	10^{16}	0.85
Polyimide aramid fiber	250	3–7	0.15	4.0	3.6	10^{12}	10^{12}	1.50
Polyimide quartz	250	6–8	0.30		4.0	10^{9}	10^{8}	0.50
Fiberglass/Teflon	75	20	0.26	0.2	2.3	10^{10}	10^{11}	1.10
Thermoplastic resin	190	25–30		3–4	10^{17}	10^{13}	NA	
Alumina-beryllia	NA	5–7, 21.0	44.0	8.0	10^{14}	10^{6}		
Aluminum (6061 T-6)	NA	23.6	200	10	NA	10^{6}		NA
Copper (CDA 101)	NA	17.3	400	17	NA	10^{6}		NA
Copper-clad Invar	NA	3–6	150XY/20Z	17–22	NA	10^{6}		NA
Copper-clad molybdenum 13%/74%/13%	NA	5.7	209XY/161Z	43.5	NA			
Copper-clad molybdenum 20%/60%/20%	NA	6.9	244XY/195Z	41	NA	10^{6}		NA
Graphite (P-100)	160	– 1.15	110XY/1.1Z	37	NA			
Graphite (P-75)	160	– 0.97	40XY/.7Z	25	NA			

NOTES: 1. These materials can be tailored to provide a wide variety of material properties based on resins, core materials, core thickness, and processing methods.

2. The X and Y expansion is controlled by the core material and only the Z axis is free to expand unrestrained. Where the Tg will be the same as the reinforced resin system used.

3. When used, a compliant layer will conform to the CTE of the base material and to the ceramic component, therefore reducing the strain between the component and P&I structure.

4. Figures are below glass transition temperature, are dependent on method of measurement and percentage of resin content.

5. NA = Not applicable

SOURCE: ANSI/IPC-SM-780, Ref. 2.

Single- and Double-Sided Printed Boards,"[9] IPC-D-319, "Rigid Single-and Double-Sided Printed Boards,"[7] or IPC-D-949, "Rigid Multilayer Printed Boards."[3]

The qualification and performance of a P&I structure should be in accordance with the appropriate requirements of IPC-RF-245, "Rigid-Flex Printed Circuit Boards,"[10] IPC-FC-250, "Single- and Double-Sided Flexible Printed Wiring with Interconnections,"[11] IPC-SD-320, Rigid Single- and Double-Sided Printed Boards,"[12] IPC-MC-324, "Metal Core Boards,"[13] or IPC-ML-950, "Rigid Multilayer Printed Boards."[14]

1.4.1 Organic base material P&I structures

Organic-base materials work best with through-hole-mounted components and surface-mounted leaded chip carriers. With leadless chip carriers, however, the thermal expansion mismatch between package and substrate can cause problems. Also, flatness, rigidity, and thermal conductivity requirements may limit their use. Finally, you must pay attention to package size, input/output (I/0) count, thermal cycling stability, maximum operating temperature, and solder joint compliance.

1.4.1.1 Epoxy-fiberglass materials. Fiberglass reinforced epoxy is widely used in conventional P&I structures featuring through-hole (Fig. 1.6) and leaded surface connections. Thermal expansion mismatch with leadless ceramic chip carriers limits usage where the I/0 count is high (above 44), where thermal or power cycling is required over large temperature extremes or a large number of cycles, or where a combination of high I/0 count and thermal cycling is required.

1.4.1.2 Polyimide-fiberglass materials. P&I structures of fiberglass-reinforced polyimide have improved heat resistance over epoxy glass, and have a slightly lower coefficient of thermal expansion. Designs should also be limited to leaded chip carriers and leadless chip carriers of relatively low I/0 count (44 or less) when thermal shock and cycling over large temperature extremes and extended life are required.

1.4.1.3 Epoxy aramid fiber materials. The coefficient of thermal expansion (CTE) of P&I structures of aramid fiber (Kevlar) reinforced epoxy closely matches that of the ceramic chip carrier, so these P&I structures are appropriate for high-I/0-count chip carrier packages and where temperature cycling over wide temperature extremes is required. Because of stresses that are generated between the fibers and resin matrix during temperature cycling, microcracks in the surface of

Figure 1.6 Typical organic-base printed circuit board assembly. *(Courtesy of Philco-Ford.)*

the substrate have been found. These microcracks are minimized with certain resin systems. Water absorption is higher than with epoxy fiberglass materials, and could cause processing problems if not removed before using. Water absorption in field environments could also be a problem unless some sealing or coating method is used.

1.4.1.4 Polyimide aramid fiber materials. Higher heat resistance makes this laminate slightly better for rework with heated tools than the epoxy materials, as well as permitting temperature cycling over greater extremes. Its low Z axis expansion also improves the reliability of through vias in the substrate. Because of its low CTE, it is ideally suited for high-I/0-count packages and where extreme environments and long life are prerequisites. However, these materials have problems with water absorption and microcracking.

1.4.1.5 Polyimide quartz materials. Quartz (fused silica) reinforced polyimide resin also has a CTE that closely matches that of the ceramic component. By varying the volume fractions of fabric and resin, the CTE can be varied. However, the amount of resin must remain below 40 percent to achieve a CTE of 6–8 PPM/°C. Slight changes in resin content and/or layers of copper drastically affect the overall CTE of the P&I structure.

Quartz reinforced materials are extremely abrasive, and thus shorten drill life. New drill design and development may ease this problem. Fabric availability is also a problem.

1.4.1.6 Fiberglass aramid fiber materials. A fiberglass-reinforced resin system can be combined with an aramid fiber–reinforced resin system to produce a low expansion composite material. The volume fraction of the aramid fiber can be adjusted to match the CTE value of the ceramic component. The aramid fiber–reinforced material may be either on internal or external layers. When the aramid fiber material is confined to the internal layers, the resin microcracking is also contained. This eliminates surface cracks that could create reliability problems such as cracked conductors, while improving the machinability and drilling of the aramid fiber containing P&I material. Its tailorable CTE makes it suited for all leadless applications.

1.4.1.7 Teflon fiberglass materials. Due to its low dielectric constant, this material is used extensively in RF applications. It is only recommended for use with small leadless components because of its high expansion coefficient. Single-sided and double-sided construction are common, whereas multilayer construction is rare.

1.4.1.8 Flexible-dielectric structures. Surface-mounted components can be used with flexible printed circuit boards. Three significant benefits resulting from this are

- Three-dimensional packaging is possible
- Stresses at the solder joint are minimized during thermal cycling
- P&I structure weight can be minimized

However, flexible printed boards will probably require the modification of some manufacturing process and will limit the usable base materials.

1.4.1.9 Thermoplastic resin P&I structures. High-performance engineering-grade thermoplastic resins are an alternative to conventional organic-base printed circuit board materials for general-purpose and surface-mount applications. Typical thermoplastics for surface mount use are polyethersulfone (PES), polysulfone (PSF), and polyetherimide (PE).

Injection-molded thermoplastic-base printed circuit boards are well suited for high-volume applications requiring a three-dimensional P&I structure with fully additive or semiadditive printed wiring. Copper-clad thermoplastic base materials are also available for planar subtractive printed wiring applications.

1.4.2 Nonorganic base materials

Nonorganic base materials typically used with thick- or thin-film technology are also ideally suited for leaded and leadless chip carrier designs. They can incorporate thick- or thin-film resistors directly on the P&I structure and buried capacitor layers that increase density and improve reliability. However, repairability of the P&I structure is limited.

Ceramic materials, usually alumina, appear ideal for P&I structure with leadless ceramic chip carriers because of their relatively high thermal conductivity (see Table 1.9) and their CTE match. Unfortunately, the P&I structure is limited to approximately 22,600 mm^2 (35 in^2). However, the evolving use of these materials with nonnoble metals, such as copper, has attracted both military and commercial applications. Ceramic P&I structures currently have two applications: ceramic hybrid circuits, and ceramic printed circuit boards. Ceramic printed circuit boards are assembled with directly soldered surface-mount components and used as independent P&I structures exclusive of a hermetic package. Ceramic boards are typically 1.0 to 1.5 mm (0.040 to 0.060 in) thick and 1000 to 22,000 mm^2 (1.5 to 35 in^2). They are usually alumina material with interconnect circuit patterns formed using multilayer thick-film or cofired processing techniques. Ceramic boards have established reliability for high-density ceramic leadless chip carrier packaging.

The boards match the CTE of a ceramic leadless chip carrier, provide a flat and rigid mounting surface, and conduct heat. Ceramic board thick-film circuits require protection from moisture absorption, so hermeticity requirements and mechanical integrity are major considerations. Typically, a glass layer is fired on as a top protective seal. Ceramic boards larger than 10,000 mm^2 (16 in^2 require mechanical support. Typically, they are bonded to a back plate for connector and chassis mounting. Select materials carefully to achieve CTE compatibility. Ceramic printed circuit boards are a new technology and standards are being developed.

1.4.3 Supporting-plane P&I structures

Supporting metallic or nonmetallic planes can be used with conventional printed circuit boards or with customer processing to enhance

P&I structure properties. Depending on the results desired, the supporting plane can be electrically functional or not and can also serve as a structure stiffener, heat sink, and/or CTE constraint.

1.4.3.1 Printed circuit boards bonded to support plane (metal or nonmetal). A conventional thin printed circuit board that has been fabricated and bonded with a rigid adhesive or insulation to a supporting plane, such as metal or graphite-fiber resin composite, can create a P&I structure with controlled thermal expansion in the x and y axes, improved rigidity, improved thermal conductivity, etc., depending on the properties of the supporting plane. However, the printed circuit board must be thin enough to preclude warping of the assembly or else the board should be bonded to both sides of the plane.

The printed circuit board portion of the P&I structure can be either unpopulated or completely assembled and tested prior to being bonded. However, components can only be mounted to one side of the printed board. Also, the support is not normally electrically connected to the printed circuit board.

1.4.3.2 Sequentially processed structures with metal support planes. High-density, sequentially processed multilayer P&I structures are available with organic dielectrics of specific thickness, ultrafine conductors, and solid-plated vias for layer-to-layer interconnections with thermal lands for heat transfer, all connected to a low-CTE metal support heat sink. Thus, this technology combines laminating materials, chemical processing, photolithography, metallurgy, and unique thermal transfer innovations in such a way that it is also appropriate for mounting and interconnecting bare integrated circuit chips.

The major advantage of this system is that the vias can be as small as 0.20 mm (0.005 in) square and conductor widths can range for 0.12 to 0.20 mm (0.003 to 0.005 in) for high interconnection density. Thus, most applications can be satisfied with two signal layers with additional layers for power and ground.

1.4.3.3 Discrete-wire structures with metal support planes. Discrete-wire P&I structures have been developed specifically for use with surface-mounted components, as shown in Fig. 1.7. These structures are usually built with a low-expansion metal support plane that also offers good heat dissipation. The interconnections are made by discrete 0.06-mm (0.0025 in) diameter insulated copper wires precisely placed on a 0.03-mm (0.0013-in) grid by numerically controlled machines. This geometry results in a low-profile interconnection pattern with excellent high-speed electrical characteristics and a density normally associated with thick-film technology.

Figure 1.7 Discrete-wire packaging and interconnecting structure assembly. *(Courtesy of Kollmorgen Corp.)*

The wiring is encapsulated in a compliant resin to absorb local stresses and dampen vibration. Electrical access to the conductors is by 0.25-mm (0.010-in) diameter copper vias. The small via size can be accommodated in the component attachment land, thus eliminating the need for fan-out patterns when using components with terminals on centers as close as 0.6 mm (0.025 in), and allowing very high packaging densities.

1.4.3.4 Flexible printed circuit boards with metal support planes. Another arrangement for a P&I structure with leadless components involves conventional fine-line polyimide flexible printed circuitry. These assemblies can be constructed in multilayer form while retaining the low-modulus feature that reduces residual strain at the solder joints. Furthermore, lasers can drill very fine holes in the thin printed circuit board laminate. These holes can be plated-through or filled with solid copper, as required.

To retain inherent flexibility while dissipating heat from the solder joint, cutouts in the flexible circuit accommodate pillars from the metal heat-sink support plane. Although this appears to be heavy and cumbersome, if the heat-sink base plates are made from thin sheets of aluminum, the resulting density of the combined circuit–heat-sink assembly might actually be less than other constructions.

1.4.4 Constraining-core P&I structures

As with supporting-plane P&I structures, one or more supporting metallic or nonmetallic planes can serve as a stiffener, heat sink, and/or CTE constraint in constraining-core P&I structures.

1.4.4.1 Porcelainized-metal (metal-core) structures. An integral core of low-expansion metal (e.g., copper-clad Invar), can reduce the CTE of porcelainized-metal P&I structures so that it closely matches the CTE of the ceramic chip carrier. Also, the P&I structure size is virtually unlimited. However, the low melting point of the porcelain requires low-firing-temperature conductor, dielectric, and resistor inks.

A number of composite P&I structures use leadless components. An integral material with a lower CTE than that of the printed circuit boards controls the CTE of these structures.

1.4.4.2 Printed circuit boards with constraining (not electrically functional) cores. Printed circuit boards bonded back-to-back to a constraining core can be used for high-density, low-warpage P&I structures. The core acts as a heat sink, but in this case is not electrically functional. For optimum density with this approach, use a multilayer construction with a centrally located, predrilled, low-CTE core. The holes in the core are filled with a compatible resin prior to lamination and the P&I structure completed with conventional fabrication techniques.

Molybdenum can be used as the core in these P&I structures for special applications that require inherent stiffness in extreme environments, but molybdenum and copper-clad Invar are difficult materials to fabricate using conventional processes. Graphite can be used where thermal conductivity per unit of weight is important.

1.4.4.3 Printed circuit boards with electrically functional constraining cores. More conventional multilayer printed circuit boards can be made as P&I structures with thin, 0.1 to 0.25 mm (0.004 to 0.010 in), copper-clad Invar as electrically functional ground and power planes.

After the planes have been predrilled they are located in a symmetrical arrangement within the lay-up and subsequently laminated as an integral part of the multilayer P&I structure. The overall CTE of the structure can be tailored by varying the composition and thickness of the planes.

1.4.4.4 Printed circuit boards with constraining cores. A constraining fiber resin composite internal plane in a conventional printed circuit board can modify thermal expansion in the x and y axes, improve rigidity, and improve thermal conductivity, depending on the properties of the supporting plane and its location within the P&I structure. These constraining fibers can be graphite, aramid fiber, quartz, etc. The very high modulus of these materials requires a balanced construction to prevent bowing or twisting.

Graphite is expensive, but its cost is justified if low weight is critical. Graphite is conductive; therefore via holes must be drilled oversize and then filled with resin prior to final via hole drilling. Graphite allows excellent CTE tailoring. Aramid and quartz fibers require modified fabrication techniques due to their mechanical properties.

1.4.4.5 Compliant-layer structures. Compliant-layer P&I structures were developed specifically for use with chip carriers and surface mounting. A novel elastomeric first layer provides cushioning and minimizes the risk of solder-joint failure due to differential thermal expansion between the structure and large surface-mounted chip carriers. Also, a combination of fine lines and spaces and either extremely small vias or plated pillars increases interconnection density. Lastly, thermal dissipation is enhanced by constructing integral metal layers for heat conduction.

1.5 Mechanical Considerations[8]

1.5.1 Bow and twist

Proper printed circuit board design, with respect to balanced circuitry distribution and component placement, is important to minimize the degree of bow and twist of the printed circuit board. Additionally, the cross-sectional layout (dielectric thickness, copper conductor density) should be kept as symmetrical as possible about the center of the board.

1.5.2 Support

Adequate mechanical support should be provided typically for at least two opposite edges of a printed circuit board assembly. The location and method of support should be such as to minimize shock and/or vibration to a level that will protect against fracturing or loosening of circuitry foil or breaking of the components or component leads as a result of flexing the printed circuit board assembly, within the tolerance of the applicable specification.

1.5.3 Structural strength

The wide variety of materials and resins available places a serious analytical responsibility on the designer when structural properties are important. The structural properties of laminates are influenced by environmental conditions that vary with the layup and composition of the base materials. Physical and electrical properties vary widely over temperature and loading ranges. The ultimate properties of printed circuit board materials are of marginal use to the designer trying to employ the printed circuit board as a structural member. The concern to meet electrical performance requirements, which are impacted by deformation and elongation of the printed circuit board, should consider lower values of ultimate material strength than those listed in the technical literature for determining structural needs.

1.5.3.1 Material selection. The first design step in the selection of a laminate, when structural strength is a major concern, is to thoroughly define the service requirements that must be met, such as environment, vibration, *g* loadings, shock (impact), and physical and electrical requirements. The choice of laminate should be made from standard structures to avoid costly and time-consuming proof-out tasks. Several laminates may be candidates, and the choice should be optimized to obtain the best balance of properties.

Materials should be easily available in the form and size required. Special laminate may be costly and have long lead times. The tolerances should be verified according to need, and the entire analysis should be checked. Items to consider are such things as machining, processing, processing costs, and the overspecification of the raw material. In addition, other items that may be important in the structural comparison of various materials are

- Resin formula
- Flame resistance
- Thermal stability
- Mechanical strength
- Electrical properties
- Flexural strength
- Maximum continuous safe operating temperature
- Reinforcing sheet material
- Nonstandard sizes and tolerances
- Machinability or punchability

- Coefficients of thermal expansion (CTE)
- Overall thickness tolerances

In addition to the above, the structural strength of the board must be able to withstand the assembly and operational stresses.

1.5.3.2 Vibration design. The design of printed circuit boards which will be subjected to vibration while in service requires that special consieration be given to the board prior to board layout. The effect on the board assembly caused by the vibration can seriously reduce the reliability of the assembly. The interrelationship between the unit, printed circuit board assemblies, and their mounting and the environmental conditions make necessary the need for a vibration analysis of the complete system very early in its design. The effect from vibration on any item within a unit can make the vibation analysis very complex.

Vibration analysis should be done on each piece of electronic hardware which contains printed circuit board assemblies. The complexity of the analysis should depend on the vibration level to which the hardware will be subjected in service. The design of the printed circuit boards will be dependent on the level of vibration transmitted to the board. Particular attention should be given to printed circuit boards subjected to random vibration.

The following criteria should be used as guidelines for determining if the level of vibration to which the boards will be subjected is a level which would require complex vibration analysis.

- The random spectral density is at or above 0.1 g^2/Hz in the frequency range of 80 to 500 Hz and an unsupported board length (or width) of greater than 76.2 mm (3 in).
- A sinusoidal vibration level at or above 3 gs at a frequency of 80 to 500 Hz.
- The board assembly will be subjected to reliability development growth testing (RDGT) at a spectral density at or above 0.07 g^2/Hz for more than 100 h in conjunction with temperature cycling.

1.5.3.3 Vibration design guidelines. The following guidelines should be observed during the design of printed circuit boards to eliminate vibration-induced failures of the printed circuit board assemblies.

- The board deflection, from vibration, should be kept below 0.08 mm/mm (0.003 in/in) of board length (or width) to avoid lead failure on DIP or similar multiple-lead devices.

- The maximum g level on the board should be kept below 100 g.

- Positive support of all components with a weight of more than 5 g (0.18 oz) should be considered when the board will be subjected to high-level vibration.

- The use of board stiffeners and/or metal cores should be considered to reduce the board deflection.

- Cushioned mounting of relays should be considered for their usage in high-level vibration environments.

- Vibration isolators should be considered for mounting of units whenever practical.

- The mounting height of freestanding components should be kept to a minimum.

- Nonaxial leaded components should be side-mounted.

Because of the interrelationship of the many components that make up a system, the use of the above guidelines does not assure the success of a unit subjected to a vibration test. A vibration test of a unit is the only way to ensure that a unit will be reliable in service.

1.5.4 Constraining-core boards

When structural, thermal, or electrical requirements dictate the use of a cored board, the board should be designed to meet the performance requirements of the applicable specification. Whether for thermal or constraining characteristics, the cored board may be designed to be symmetrical or unsymmetrical. There are some advantages in an unsymmetrical design in that the electrical properties or functions are separated from the mechanical or heat-dissipation functions. The analysis of these properties can be evaluated separately.

The drawback of the unsymmetrical design is that due to the differences of the coefficient of thermal expansion of the printed circuit board and the core material, the completed board may warp during soldering or reflow operations. Some compensation can be achieved by having an additional copper plane added to the back of the interconnection product. The extra copper plane increases the expansion coefficient slightly, but a positive effect is that it enhances thermal conductivity.

A more desirable construction may be that of the symmetrical cored board. In those configurations equal amounts of printed circuit boards are laminated to either side of the metal or constraining core. This concept provides excellent properties for assembly. However, the manufacturing steps are more complex, especially if through connections must be made between side 1 and side 2.

1.5.5   Mechanical hardware attachment

The printed circuit board should be designed in such a manner that mechanical hardware can be easily attached, either prior to the main component insertion, or subsequently. Sufficient physical and electrical clearance should be provided for all mechanical hardware that requires electrical isolation.

1.5.6   Part support

All parts weighing 7 g (0.25 oz) or more per lead should be supported by specified means which will help ensure that their soldered joints and leads are not relied upon for mechanical strength.

- The worst-case levels of shock and vibration environment for the entire structure in which the printed circuit board assembly resides, and the ultimate level of this environment that is actually transmitted to the components on the board. (Particular attention should be given to equipment that will be subjected to random vibration.)

- The method of mounting the board in the equipment to reduce the effects of the shock and vibration environment, specifically the number of mounting supports, their interval, and their complexity.

- The attention given to the mechanical design of the board, specifically its size, shape, type of material, material thickness, and the degree of resistance to bowing and flexing that the design provides.

- The shape, mass, and location of the components mounted on the board.

- The component lead wire stress-relief design as provided by its package, lead spacing, lead bending, or a combination of these, plus the addition of restraining devices.

- The attention paid to workmanship during board assembly, so as to ensure that component leads are properly bent, not nicked, and that the components are installed in a manner that tends to minimize component movement.

- Conformal coating may also be used to reduce the effect of shock and vibration on the board assembly.

1.5.7   Thermal management

An increasing variety of integrated circuit packages and multilayer interconnection technologies are currently being developed in the electronics industry to capitalize on integrated circuit density improvements. As components become larger and more complex, heat dissipation will increase. There is little point in using denser compo-

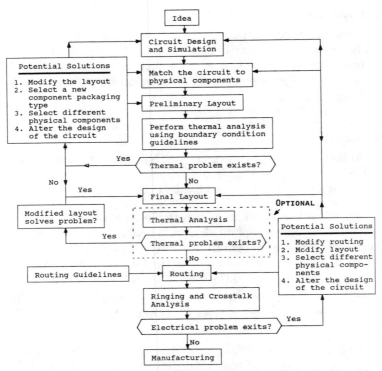

Figure 1.8 Thermal design methodology. *(Courtesy of Pacific Numerix Corp.)*

nents if they must be widely spaced on the interconnect structure to dissipate heat. The primary aim of electronic thermal control is to prevent thermally induced total loss of a component's electronic function.

The methodology for optimizing the thermal (and electrical) design of a printed circuit board is shown in Fig. 1.8. First, the temperature of all components must be maintained within functional allowable limits. A functional temperature limit is the temperature range within which the electrical circuit should operate. Second, the distribution of component operating temperatures must not cause excessive system failure.

In addition to these primary objectives, there are several secondary objections:

- The overall cooling subsystem meets availability and reliability objectives.

- The overall cooling system design is consistent with the customer's environment and heat-rejection capability.

- Exposed surfaces are kept within safety limits.
- The cooling system is consistent with the overall machine cost.

Keep all of these objectives in mind throughout the system design process. As in many engineering designs, the optimum system requires technical trade-offs.

Several factors must be considered when selecting the cooling hardware. Military systems for field deployment are most often indirectly cooled, whereas test equipment and commercial systems continue to use direct cooling air impingement on components. For stationary ground equipment, the weight of a cooling system is not a major issue, but it is extremely important in portable, avionic, and spacecraft equipment. The cost, reliability, energy consumption, maintainability, and repairability of the cooling hardware affects the cost of the entire system.

1.6   Electrical Considerations[15]

1.6.1   Conductor thickness and width

The minimum width and thickness of conductors on the finished board should be determined on the basis of the current-carrying capacity required, and the maximum permissible conductor temperature rise. The minimum conductor width and thickness should be in accordance with Fig. 1.9 for conductors on external and internal layers of the printed circuit board. The following list discusses the use of Fig. 1.9.

- The design chart has been prepared as an aid in estimating temperature rises (above ambient) versus current for various cross-sectional areas of etched copper conductors. It is assumed that for normal design, conditions prevail where the conductor surface area is relatively small compared to the adjacent free panel area, the curves as presented include a nominal 10 percent derating (on a current basis) to allow for normal variations in etching techniques, copper thickness, conductor width estimates, and cross-sectional area.

- Additional derating of 15 percent (currentwise) is suggested under the following conditions: (a) For panel thickness of $\frac{1}{32}$ in or less. (b) For conductor thickness of 0.0042 in (3 oz/ft^2) or thicker.

- For general use the permissible temperature rise is defined as the difference between he maximum safe operating temperature of the laminate and the maximum ambient temperature in the location where the panel will be used.

- For single-conductor applications, the chart may be used directly

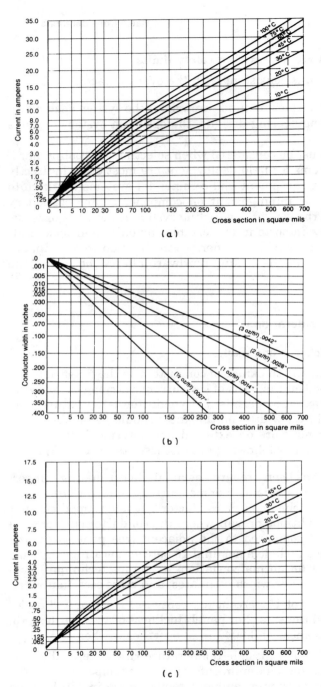

Figure 1.9 Conductor thickness and width for internal and external printed circuit board layers. *(From ANSI/IPC-D-949).[9]*

for determining conductor widths, conductor thickness, cross-sectional area, and current-carrying capacity for various temperature rises.

- For groups of similar parallel conductors, if closely spaced, the temperature rise may be found by using an equivalent cross-section and an equivalent current. The equivalent cross-section is equal to the sum of the cross-section of the parallel conductors, and the equivalent current is the sum of the currents in the conductors.

- The effect of heating due to attachment of power-dissipating parts is not included.

- The conductor thicknesses in the design chart do not include conductor overplating with metals other than copper.

For ease of manufacture and durability in usage, these parameters should be optimized while maintaining the minimum recommended spacing requirements. To maintain finished conductor widths as on the master drawing, conductor widths on the production master may require compensation for process allowances.

1.6.2 Electrical clearance

Spacings between conductors on individual layers should be maximized whenever possible. The minimum spacing between conductors, between conductive patterns, and between conductive materials (such as conductive markings or mounting hardware) and conductors should be in accordance with Table 1.1 and defined on the master drawing. Conductive markings may touch a conductor on one side, but minimum spacing between the character or marking and adjacent conductors should be maintained.

To maintain the conductor spacing shown on the master drawing, space widths on the production master may require compensation for process allowances. Plated-through holes passing through internal foil planes (ground and voltage) and thermal planes should meet the same minimum clearance between the plated-through hole and foil or ground planes as required for spacing between internal conductors.

1.6.3 Electrical performance

Printed circuit boards can often be required to satisfy a wide range of electrical performance needs. Such requirements may affect the board's construction and material selection, especially when transmission-line characteristics are necessary. Chapter 2 deals with many of these design considerations.

1.7 Design for Producibility

Printed circuit board manufacturing costs can be significantly reduced by taking manufacturability into account early in the design cycle. Board real estate use, board size and shape, laminate material, and panel size are design choices that affect manufacturability.

Conductor spreading improves board manufacturability by closely centering copper between lands and vias, straightening "stair-stepped" conductors, and eliminating unnecessary sharp bends. Software packages maximize clearance between conductors, lands, and vias, reduce the number of vias, and provide optimal conductor routing. High-density digital and analog circuitry performance also benefit from implementing designs this way. These design software routines can automatically chamfer conductor bends at 45° angles. User-specified parameters control minimum and maximum chamfer length and escapes for lands and vias. Corners can also be rounded automatically. This feature shortens the overall length of conductors and improves the electrical characteristics of the circuit.

1.7.1 Panelization

Board size and shape have the greatest effect on printed circuit board cost. Designers are not always free to choose board dimensions, however. Those parameters are sometimes dictated by the needs of the product being built. An understanding of the principles involved and an overview of manufacturing requirements are provided to designers in IPC-D-322 "Guidelines for Selecting Printed Wiring Board Sizes Using Standard Panel Sizes."[16]

The largest panel size that can be processed economically is a function of sheet laminate common in the marketplace. This dictates that recommended guidelines for board fabrication panel sizes should be submultiples of common sheet sizes. Other choices lead to cost penalties in wasted material due to excess trim. Restructions in material use apply equally to the fabricator using mass lamination techniques, either in-house or at a laminate supplier. Some board size recommendations for optimum panel utilization are shown in Table 1.10.

On dense circuits, where dimensional stability is important, the grain direction affects end-product board stability. The long dimension of the finished board should be parallel with the fill direction of the laminate sheet. Factors of equipment, material, and human engineering do not accommodate the use of these full-size sheets by the board fabricator. Therefore, a smaller fabrication panel size that is the largest submultiple of the full-size sheet is recommended as the primary standard panel size.

Secondary standard panel sizes should be submultiples of the full-

TABLE 1.10 Printed Circuit Board Size Recommendations as a Function of Panel Size and Material Utilization

A	B	C	D	E	F	G
Panel size, mm (in)	Panel area, cm² (in²)	Quantity of boards per panel	Board size, mm (in)	Board area, cm² (in²)	Material Utilization $\frac{C \times E \times 100}{B}$, %	Notes & Restrictions
457 × 610 (18 × 24)	2788 (432²)	1	406 × 559 (16 × 22)	2270 (352²)	81	1, 3, 4, 5, 6
457 × 533 (18 × 21)	2436 (378²)	1	406 × 483 (16 × 19)	1961 (304²)	80%	1, 3, 4, 5, 6
457 × 610 (18 × 24)	2788 (432²)	2	267 × 406 (10.5 × 16)	1084 (168²)	78	1, 2
457 × 610 (18 × 24)	2788 (432²)	3	178 × 406 (7 × 16)	723 (112²)	78	1, 2
457 × 533 (18 × 21)	2436 (378²)	2	234 × 406 (9.2 × 16)	950 (148²)	78	1, 2
406 × 457 (16 × 18)	1855 (288²)	1	356 × 406 (14 × 16)	1445 (224²)	78	1
457 × 533 (18 × 21)	2436 (378²)	3	152 × 406 (6 × 16)	617 (96²)	76	1, 2
406 × 457 (16 × 18)	1855 (288²)	2	196 × 356 (7.7 × 14)	698 (109²)	76	1, 2
356 × 457 (14 × 18)	1627 (252²)	1	305 × 406 (12 × 16)	1238 (192²)	76	1
356 × 457 (14 × 18)	1627 (252²)	2	196 × 305 (7.7 × 12)	598 (93²)	74	1, 2
305 × 457 (12 × 18)	1394 (216²)	1	254 × 406 (10 × 16)	1031 (160²)	74	1
406 × 457 (16 × 18)	1855 (288²)	3	127 × 356 (5 × 14)	452 (70²)	73	1, 2
305 × 406 (12 × 16)	1238 (192²)	1	254 × 356 (10 × 14)	904 (140²)	73	1

NOTES: 1. Using 25.4-mm (1-in) borders.
2. Using 12.7-mm (0.5-in) margins for test coupon (one per panel) and 0.25 in for margins without test coupon rounding off to the next smallest 2.5-mm (0.1-in) increment.
3. Exceeds normal autoinsertion limits.
4. Exceeds normal in-circuit test limits.
5. Exceeds normal bare-board test limits.
6. Exceeds normal solder coating limits.
SOURCE: ANSI/IPC-D-322, Ref. 16.

sheet size, not of the primary standard. The larger sizes are typically the most effective from the standpoint of labor cost per unit area processed, but may pose difficulties to the board fabricator.

1.7.2 Bare board fabrication

The limitations of equipment and people must be considered when attempting to maximize board manufacturability and reduce costs.

Without mechanical assists, which are absent in most board fabrication facilities, full-size laminate sheets are beyond the capability of human limits of weight, reach, and control. In order to avoid unusual yield problems due to edge effects of plating and to allow multiple-image processing, borders and margins are commonly employed by the board fabricator. Such borders usually range in size from 9.5 to 3.8 mm (0.375 to 1.5 in). Size is determined by circuit density and must accommodate racking for plating.

Margins between boards range from 2.5 to 1.3 mm (0.1 to 0.50 in); but are usually 2.5 mm (0.10 in) for blanking, 6.3 mm (0.25 in) for routing, and 1.3 mm (0.50 in) for shearing (or the nearest dimension to maintain grid relationships). Space allocated for test coupons and tooling holes must be taken into account when determining the areas on the fabrication panel that remain for use by the printed circuit boards.

Using the largest possible board size minimizes costs and maximizes production efficiency. Sizes are most often determined, however, by the area required to contain the desired electrical functions.

Panelization, editing, and transfer workstations are making printed circuit board manufacturing more productive. These systems are designed for fast, easy editing, and allow individual, type group, or global edits that are selectable by windows and delimiters. Specific design enhancements that can be made include.

- Adjusting conductor width and Spacing
- Changing land size and shape
- Rerouting circuits
- Adding jumpers
- Resizing x and y axes independently
- Stretching geometric figures
- Removing unused lands

These are just a few of the design enhancements that can improve printed circuit board producibility and reduce board costs. Other benefits that can be expected from a smoothly operating design for manufacturability system include:

- Products that are developed on schedule
- Products that meet cost targets for optimum profitability
- Reliability improvements that equal or exceed product targets
- Reductions in the number of parts and part numbers needed

- Improvements in product quality
- Products that meet all specifications
- Development of better products with significantly reduced time to market

1.7.3 Achieving optimum results

Designing for manufacturability is the way efficiently managed electronic companies organize their product development resources. They use the best design and manufacturing tools and most optimum operational practices. However, it is not only tools, checkpoint reviews, and use of the best practices that lead to success; it is much, more more. It is all of the elements needed to design and manufacture electronic products in order to obtain a competitive advantage. In order to achieve this goal it is essential to establish a multifunctional development team that coordinates the efforts of the personnel responsible for

- Electrical design
- Mechanical design
- Manufacturing engineering
- Purchasing
- Production engineering
- Quality engineering
- Service engineering
- Reliability engineering
- Materials engineering
- Printed circuit board fabrication
- Printed circuit board assembly
- Production management
- Product marketing
- Key supplier services

Although there are many approaches being taken to design for manufacturability, most good programs have several objectives in common. Meeting customer requirements and quality expectations are two common goals. Another is tailoring manufacturing processes to make products at the lowest competitive cost and highest quality. Reducing product development cycles to provide products to the market in less time is yet another shared objective. The best results are achieved when the designer and the board manufacturer discuss objectives early in the design cycle.

1.8 Design for Testability[3,7]

Normally, prior to starting a design, a testability review meeting should be held. Testability concerns, such as circuit visibility, circuit operation, and special test requirements and specifications are discussed as a part of the test strategy. Any artwork or configuration concepts necessary to test the board should be incorporated into the final board layout.

During the design testability review meeting, tooling concepts are established, and determinations are made as to the most effective tool cost versus board layout concept conditions.

During the layout process, any printed circuit board changes that impact the test program, or the test tooling, should immediately be reported to the proper individuals for determination as to the best compromise. The testing concept should develop approaches that can check the board for problems and also detect fault locations wherever possible.

The following provides a checklist to be used in evaluating the testability of the design.

- Route test and control points to the edge connector to enable monitoring and driving of internal board functions and to assist in fault diagnosis.
- Divide complex logic functions into smaller, combinational logic sections.
- Avoid one-shots; if used, route their signals to the edge connector.
- Avoid potentiometers and "select-on-test" components.
- Use a single, large-edge connector provide input/output (I/0) pins and test and control points.
- Make printed circuit board I/0 signals TTL-compatible to keep automatic test equipment (ATE) interface costs low and give flexibility.
- Provide adequate decoupling at the board edge and locally at each integrated circuit.
- Provide signals leaving the board with maximum fanout drive, or buffer them.
- Buffer edge-sensitive components from the edge connector—such as clock lines and flip-flop outputs.

- Do not tie outputs together.

- Never exceed the logic rated fan-out; in fact, keep it to a minimum.

- Do not use high-fan-out logic devices. Do use multiple fan-out devices, and keep their outputs separate.

- Keep logic depth on any board to a low level by using edge-terminated test and control points.

- Single-load each signal entering the board whenever possible.

- Terminate unused logic pins with a resistive pull-up to minimize noise pickup.

- Do not terminate logic outputs directly into transistor bases. Do use a series current-limiting resistor.

- Buffer flip-flop output signals before they leave the board.

- Use open-collector devices with pull-up resistors to enable external override control.

- Avoid using redundant logic to minimize undetectable faults.

- Bring outputs of cascaded counters to higher-order counters so that they can be tested without large counts.

- Construct trees to check the parity of selected groups of 8 bits or fewer.

- Avoid wire-OR and wire-AND connections. If you cannot, use gates from the same IC package.

- Provide some way to bypass level-changing diodes in series with logic outputs.

- Break paths when logic element fans out to several places that converge later.

- Use elements in the same IC package when designing a series of inverters or inverters following a gate function.

- Standardize power-on and ground pins to avoid test-harness multiplicity.

- Bring out test points as near to d/a conversions as possible.

- Provide a means of disabling on-board clocks so that the tester clock may be substituted.

- Provide mounted switches and resistor-capacitor networks with override lines to the edge-board connector.

- Route logic drives of lamps and displays to the edge connector so that the tester can check for correct operation.

- Divide large printed circuit boards into subsections whenever possible.
- Separate analog circuits from digital logic, except for timing circuits.
- Uniformly mount ICs and clearly identify them to make it easier to locate them.
- Provide sufficient clearance around IC sockets and direct-soldered ICs so that IC clips can be attached whenever necessary.
- Add top-hat connector pins or mount extra IC sockets when there are not enough edge-board-connector pins for test and control points.
- Use sockets with complex ICs and long, dynamic shift registers.
- Wire feedback lines and other complex circuit lines to an IC socket with a jumper plug so that they can be interrupted at test.
- Use jumpers that can be cut during debugging. The jumpers can be located near the edge-board connector.
- Fix locations of power and ground lines for uniformity among several board types.
- Make the ground trace large enough to avoid noise problems.
- Group together signal lines of particular families.
- Clearly label all parts, pins, and connectors.

References

1. "Automated Design Guidelines," ANSI/IPC-D-390, Revision A, February 1988, Institute for Interconnecting and Packaging Electronic Circuits, Lincolnwood, Ill.
2. "Component Packaging and Interconnecting with Emphasis on Surface Mounting", ANSI/IPC-SM-780, July 1988, Institute for Interconnecting and Packaging Electronic Circuits, Lincolnwood, Ill.
3. "Design Standard for Rigid Multilayer Printed Boards", ANSI/IPC-D-949, January 1987, Institute for Interconnecting and Packaging Electronic Circuits, Lincolnwood, Ill.
4. "Suggested Guidelines for Modification, Rework and Repair of Printed Boards and Assemblies", ANSI/IPC-R-700, Revision C, January 1988, Institute for Interconnecting and Packaging Electronic Circuits, Lincolnwood, Ill.
5. Gary Ferrari, Custom Photo and Design Inc., "Printed Wiring Board Standards," *Circuit Design*, December 1989, pp. 27–38.
6. "Design Standard for Rigid Printed Boards and Rigid Printed Board Assemblies," IPC-D-275, September 1990, Institute for Interconnecting and Packaging Electronic Circuits, Lincolnwood, Ill.
7. "Design Standard for Rigid Single- and Double-Sided Printed Boards," ANSI/IPC-D-319, January 1987, Institute for Interconnecting and Packaging Electronic Circuits, Lincolnwood, Ill.
8. Gerald L. Ginsberg, Magnavox Corp., "Basic Printed-Wiring Design Considerations," *Electronic Packaging & Production*, July 1974.

9. "Design Standard for Flexible Single- and Double-Sided Printed Boards," ANSI/IPC-D-249, January 1987, Institute for Interconnecting and Packaging Electronic Circuits, Lincolnwood, Ill.

10. "Performance Specification for Rigid-Flex Printed Boards," ANSI/IPC-RF-245, April 1987, Institute for Interconnecting and Packaging Electronic Circuits, Lincolnwood, Ill.

11. "Specification for Single- and Double-Sided Flexible Printed Wiring," ANSI/IPC-FC-250, Revision A, September 1986, Institute for Interconnecting and Packaging Electronic Circuits, Lincolnwood, Ill.

12. "Performance Specification for Rigid Single- and Double-Sided Printed Boards," ANSI/IPC-SD-320, Revision B, November 1986, Institute for Interconnecting and Packaging Electronic Circuits, Lincolnwood, Ill.

13. "Performance Specification for Metal Core Boards," ANSI/IPC-MC-324, October 1988, Institute for Interconnecting and Packaging Electronic Circuits, Lincolnwood, Ill.

14. "Performance Specification for Rigid Multilayer Printed, ANSI/IPC-ML-950, Revision C, November 1986, Institute for Interconnecting and Packaging Electronic Circuits, Lincolnwood, Ill.

15. Gerald L. Ginsberg, Component Data Associates Inc., "Board Design for Optimized Fabrication," *Electronic Packaging & Production*, June 1989, pp. 46–48.

16. "Guidelines for Selecting Printed Wiring Board Sizes Using Standard Panel Sizes," ANSI/IPC-D-322, September 1984, Institute for Interconnecting and Packaging Electronic Circuits, Lincolnwood, Ill.

Electrical Performance Considerations

2.1 Introduction[1]

The packaging of electronic equipment has traditionally been considered with mechanical. An assortment of active and passive devices needed to be provided with adequate physical support, environmental protection, heat removal, electrical interconnections, and electrical insulation in a cost-effective way. However, packaging design is becoming more complex. Switching devices typical of digital electronics technology are available in increasingly greater degrees of both switching speed and count of devices per integrated circuit (IC). Individual ICs are being provided with greater numbers of connections in smaller individual package sizes. Also, the competitive need to take maximum advantage of device density and speed has forced packaging designers to pay much more attention to problems of electromagnetic wave-propagation phenomena associated with transmission of switching signals within the system.

2.1.1 Decision making

Therefore, it has become necessary to determine at the start of a project whether the packaging concerns need early introduction to the design process. Performance of older systems was limited by the devices available and packaging design could be left for the end of the design activity. Since high-performance systems are limited in speed by the interconnection wiring, it is important that the design get the most out of the printed circuit board design.

There are numerous design options to balance against the mechanical and electrical requirements of the end product (Fig. 2.1). For a

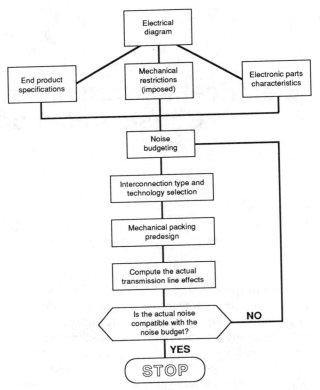

Figure 2.1 High-speed packaging design concept. *(From IPC-D-317.[1])*

proposed system a choice among types of devices has to be made. This will have an influence on the kinds of materials to be selected. It will become apparent that the design is a collection of compromises to get the best overall performance. The decision-making cycle will consist of proposing a design, with selection of devices and materials and an interconnection scheme. The system preferably will be modeled and its deficiencies and limitations identified. This should result in some alterations in the design as the start of another modeling cycle.

2.1.2 System considerations

The decision on the number of signal layers in multilayer printed circuit boards will be influenced by the degree of crossover of connections and the density of interconnections within the board. There will be compromise of degree of isolation of one interconnection from another versus the need to get a large number of interconnections in a given printed circuit board area.

Power requirements of the devices will need to be met by an adequate scheme. Often the choice is made to provide voltage planes in multilayer boards that serve the dual roles of ground plane for signal traces and leveled supply of voltage to devices. An alternative is to provide a bus system to the array of ICs on the printed circuit board.

2.2 Digital System Interactions[2,3]

There is a tendency among digital designers to simply consider digital system operation as a DC process. In the days of TTL logic, the analog effects of the chip-to-chip circuit card interconnections were almost insignificant. However, in today's complex electronic equipment, the high-speed logic families which are required to increase throughput transform printed circuit board interconnections into transmission lines.

Digital designers must now become familiar with a host of analog effects in order to cope with this new technology. Transmission line stubs, interlayer vias, voltage mismatch reflections, conductor geometry, and printed circuit board dielectric effects are some of the many factors which now demand attention early in the system design cycle.

Design of high-performance digital systems requires a match between system architecture, speed, and the interconnect and packaging integration level. To properly design a high-speed interconnect and packaging medium, engineers must be aware of a host of interdisciplinary problems and the relationships between them. In order to optimize interconnect performance and maximize reliability, the designer must as a minimum address the following goals:

- Minimize voltage mismatch reflections (ringing) created when high-edge-speed signals propagate through impedance discontinuities (vias, package pins and lands, stub junctions, edge-board connectors, unterminated loads, etc.).

- Reduce crosstalk between neighboring signal lines by maximizing signal-to-signal spacing and by minimizing signal-to-ground distances.

- Reduce chip-to-chip interconnect delay in proportion to the increase in internal IC speed by employing very short traces between devices. To increase signal velocity, low-dielectric-constant material should be used.

- Minimize signal edge-speed degradation and attenuation created by high interconnect dc resistance, skin effects, and dielectric loss effects at gigahertz frequencies.

- Minimize power and ground noise by providing low-impedance power distribution networks with ample high-frequency decoupling capacitors. Employ multiple power and ground planes.

- Ensure proper voltage transitions by fine-tuning the interconnect impedance and employing terminations. Improper impedance selection will result in voltage transitions which are insufficient to create a logic level transition, while lack of proper termination can cause excessive signal swings and false switching.

- Minimize capacitive loading on signal traces to maintain the highest characteristic impedance and shortest propagation delay.

- Account for material and manufacturing effects in the design to meet reliability and performance goals.

- Minimize thermal effects on printed circuit board dielectric and interconnect performance.

- Minimize multilayer board thickness to simplify manufacturing and heat transfer.

These factors, plus a host of others, contain many interdependencies and often are at odds with one another. For example, high wiring density is required to minimize interconnect delays. However, as signals are placed closer together, mutual coupling increases, with a corresponding rise in crosstalk levels.

At the same time, power considerations would lead toward a high characteristic impedance value to minimize termination power; however, this heightens the conductor's susceptibility to crosstalk and may make the system unreliable. Thus as system complexity increases (Fig. 2.2), there is a growing need for a more automated approach to the design cycle. It would be close to impossible to determine how the many possible printed circuit board effects (voltage mismatch reflections, crosstalk, conductor cross-section variations, stubs, etc.) would superimpose in time without the aid of analog models and computer-aided engineering (CAE) software tools. Such detailed analysis is essential if an accurate system noise margin analysis is to be performed.

The use of modeling and analysis tools must be tempered by the realities of time and cost for large simulation runs. Analysis of high-bandwidth interconnect configurations results in models which contain thousands of discrete components and require several hours for execution using accelerated workstations. Thus, all signals in a high-speed design cannot be simulated to the full extent. Hence, developed

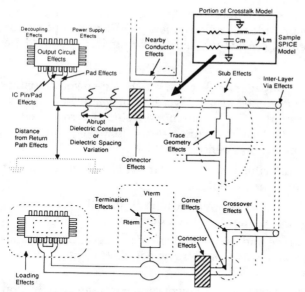

Figure 2.2 Complex interconnect structures. *(Courtesy Raytheon Co.)*

tools must be coupled with design guidelines and methodologies to handle the thousands of interconnections in a complex printed circuit board design.

The developed guidelines should address the identification of key parasitics that impact system performance and should recommend methods to improve system performance and improve performance based upon analytical, empirical, and simulation results. Central to the need for design guidelines is the importance of impedance control, crosstalk reduction, and noise budget analysis in printed circuit board design.

The desirable features for the integrated system would include the ability to perform parasitic effects analysis and prediction, modeling, and simulation of candidate physical geometries, prelayout analysis of electrical and thermal effects, high-bandwidth placement and routing, postroute analysis and simulation through circuit effects extraction, noise margin analysis, and finally postroute linkages to automated manufacturing equipment.

2.3 Power Distribution[1,3]

A predominantly important factor that should be considered in the design of printed circuit boards is power distribution. The grounding scheme, as a part of the distribution system, provides not only a dc power return but also provides a radiofrequency (RF) return plane for digital logic to be referenced. Thus, the following items should be taken into consideration:

- Maintain a lower RF impedance throughout the dc power distribution. [An improperly designed ground system can result in RF emissions as a result of radiated field gradients developed across the uneven board impedance. Also, it is unable to efficiently reduce the board's electromagnetic interference (EMI).]

- Decouple the power distribution at the printed circuit board using a 0.1- to 10.0-μF tantalum capacitor.

- Connect a ceramic capacitor, of approximately 0.1-μF, across V_{CC}; one capacitor for no more than five logic devices.

- Minimize the impedance and radiation loop of the decoupling capacitor by keeping capacitor leads as short as possible; locate them adjacent to the integrated circuit.

- Use planes wherever possible for power and voltage distribution techniques.

- When using power conductors, run power traces as close as possible to the ground traces.

- Maintain both power and ground traces as wide as possible.

- Keep power and ground planes next to each other. (They virtually become one plane at high frequencies.)

Figure 2.3 shows a bad power distribution layout with high inductances and few signal return paths; this leads to crosstalk. Figure 2.4 is a better layout, as it reduces power distribution and and logic-

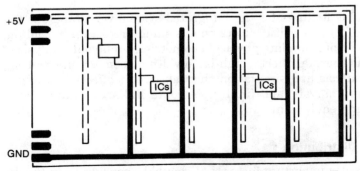

Figure 2.3 Poor voltage/ground distribution layout concept. *(From ANSI/IPC-D-319.[3])*

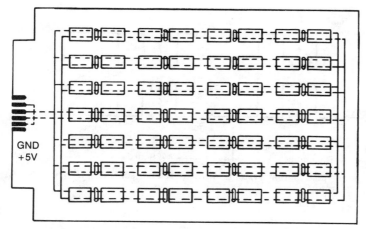

Figure   2.4 Acceptable   voltage/ground   distribution   layout concept.   *(From ANSI/IPC-D-319.[3])*

return impedances, crosstalk, and board radiation. The best layout is shown in Figure 2.5, which further reduces EMI problems.

The power distribution planes utilized in printed circuit boards do not have zero impedance. Likewise the power supply does not have a zero source resistance. A real-life model of a multilayer power distribution network is shown in Fig. 2.6. The power supply is represented by a voltage source V_s, R_{sp}, and R_{sg}. The distribution impedances are broken out as backplane and plug-in printed circuit board sheet inductances, resistances, and plane-plane capacitance.

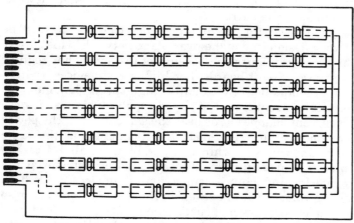

Figure   2.5 Preferred   voltage/ground   distribution   layout concept.   *(From ANSI/IPC-D-319.[3])*

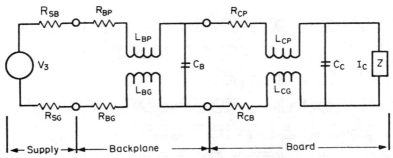

Figure 2.6 DC power distribution systems, without remote sensing. *(From IPC-D-317.[1])*

The power distribution AC impedance is subdivided into three components. The first is the switching transient impedance Z_{sw}. This impedance is between the decoupling capacitor and the V_{CC} leg of the integrated circuit. This is the highest-frequency component of the current. The second component is the impedance due to the bulk capacitor charging the IC decoupling capacitors Z_{bc}. The current in this impedance is lower in frequency and higher in amplitude than the current in the first component. The voltage drop due to the lower impedance because of the lower frequency involved will be less than the above case. The bulk decoupling capacitance refresh component Z_{rc} is the final element of the printed circuit board decoupling impedance. This current is responsible for recharging the bulk capacitors. It is supplied from the power supply and will usually have the lowest-frequency component of the current.

Each of the printed circuit board impedance components can be modeled as a transmission line plane-over-plane network. This closely models the worst-case configuration because of the somewhat regular layout of most printed circuit boards. For high-speed devices, switching activity is accompanied by equally high-speed demands for changes in electrical current from the power supply. If several devices are demanding current changes at or near the same instant, the power distribution system is required to meet these demands at the same time that it maintains voltage within specified limits to all devices being supplied. Meeting this requirement leads to power distribution designs that provide low-inductance connections to devices, with high capacitance among various voltage levels in the distribution system so that it is able to meet some level of current demand without feeding a noise signal back through the distribution network to interfere with the performance of other components.

2.4   Signal Distribution[4,5]

Interconnections and the packaging of electronic components primarily have been in the domain of designers who were concerned with interconnections specified in wire listings or to/from listings. Conductors of electrical signals were routed with only two concerns:

1. That continuity was maintained between points

2. That no shorts were permitted

Aside from providing a good electrical path, the electrical properties of the signal were not a major concern. However, with the advent of high-speed circuits imposing high-density interconnecting requirements, all of this is changing.

Even if it were possible to make a circuit capable of switching at infinite speed, it would be the interconnection that would dictate the performance of the systems using these devices. To illustrate this point, Fig. 2.7 plots the system speed of a 1-GHz gallium arsenide (GaAs) device versus interconnection path length in various dielectric materials.[6]

For a 76-mm (3-in) interconnecting path length, the speed has been reduced 25 percent for PTFE Teflon (i.e., 750-MHz), 30 percent for polyimide, and over 45 percent for ceramic. As the path lengths get longer, the degradation becomes even worse. Consequently, the interconnection's total length must be kept very short in order to preserve most of the inherent switching speeds of the newer-generation devices.

Interconnection wiring, such as a printed circuit board, has inductance, capacitance, and resistance and presents transmission line environments with built-in characteristic impedance, propagation delay, and signal line reflections. Each of these parameters is sensitive to high frequency and must be considered in the system design—the higher the frequency, the more critical the design.

In high-speed systems, two device characteristics, logic rise time and logic propagation delay, are items of importance. With the fast rise time of high-speed devices (Table 2.1), along with the shorter propagation delays, special considerations need to be given to the printed circuit board design in order to avoid degrading the performance of the system. The items of important are:

- Interconnect delay
- Line ringing
- Crosstalk

Interconnect delay is affected by length of signal path (device to pack-

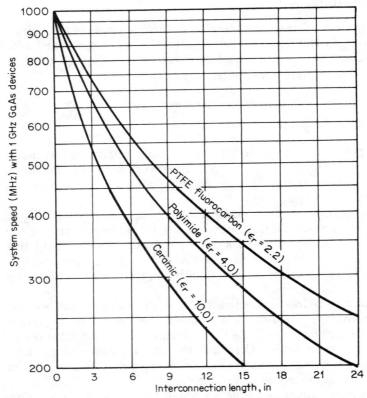

Figure 2.7 Switching speed versus interconnection length. *(Courtesy of Hughes Aircraft Co.)*

TABLE 2.1   Relationship between Frequency and Rise Time

Frequency, MHz	Rise time, ns
10	35
50	7
100	3.5
200	1.75
400	0.88
800	0.44

SOURCE: Technology Seminars Inc.

age, package to package, and board to board), material dielectric, and signal configuration. The objective is to keep the gate-to-gate interconnect delay below the device rise time (Table 2.2). When twice the propagation delay is less than the rise time of the device, transmission-line characteristics do not prevail. Conversely, when signal lengths for various logic families exceed the values in Table 2.3, transmission line parameters prevail.

For example, high-speed emitter-coupled logic (ECL) has a 0.45 ns rise time. This translates to a wavelength in air of about 380 mm (15 in), 190 mm (7.5-in) in printed circuit board, or less than 100 mm (4-in) in ceramic. This means that for printed circuit boards, if the interconnection path is more than 190 mm (7.5 in), the electromagnetic properties of the signal and the waveguide of transmission line effects should be considered.

2.4.1 Transmission lines[1,5,7]

Signal routing on printed circuit boards is sometimes done randomly with little regard to electrical considerations. This is adequate as long as electrical energy is confined to the flow of charges in the conductor without regard to its wave property. At higher frequencies the design rules change, and the signal is propagated in waves between conductors. The spacing of the dielectrics and its material characteristics assume new importance; waveguides are being designed.

At low frequencies, a signal path on a printed circuit board may usually be represented electrically as a capacitance in parallel with a resistance. However, as the frequency is increased, this approach of lumped circuit modeling breaks down, and signal paths must be regarded as transmission lines. For transmission line interconnects, the electrical and dielectric properties of the printed circuit board materials have an enhanced importance and greater care must be taken with the design and termination of the circuit.

Several attempts have been made to define the point at which conductors act as transmission lines, the required analysis being performed in either the frequency or the time domain. However, the critical point to remember for digital signals is that it is the pulse rise time, and not the rate at which the device is clocked, that is a key determining factor. For digital circuitry, if the circuit path length is greater than one-seventh of the wavelength, then the path typically must be considered a transmission line. (For analog circuitry, which is less tolerant of noise, the critical length is usually shortened to one-fifteenth of the wavelength in the dielectric medium).

An alternative equivalent definition of the point at which a printed circuit board significantly affects the transmission characteristics of a

TABLE 2.2 Typical Characteristics of Various Logic Families

Logic families	Typical output voltage swing, V	Rise/fall time, ns	Bandwidth, MHz, $1/\pi\tau_r$	Max V_{CC} voltage drop, V	Power supply transition current, mA	PS decoupling* capacitor, pF	PS current‡ per gate drive, mA	Input C, pF	DC noise† margin, mV
Emitter-coupled logic (ECL-10K)	0.8	2/2	160	0.2	1	350	1.2	3	125
Emitter-coupled logic (ECL-100K)	0.8	0.75	420	0.2	100				
Advanced Schottky	3	1.75	182	0.5	5	300			
Transistor-transistor logic (TTL)	3.4	10	32	0.5	16	2350	1.5	5	400
Low-Power TTL (LP-TTL)	3.5	20/10	21		8	400	1.6	5	400
Schottky TTL logic (STTL)	3.4	3/2.5	120	0.5	30	1500	4	4	300
Low-power schottky (LS-TTL)	3.4	10/6	40	0.25	8	3700	2.1	6	300
Complementary metal oxide logic (CMOS) 5 V or (15 V)	5 (15)	90/100 (50)	3 (6)		1 (10)	—§	0.2	5	1 V (4.5)
High-Speed CMOS 5 V	5	10	32	2	10	400	1	5	1 V

*C = (1 driving gate & 1 for 5 driven gates)•rise time/(0.2•max $V_{cc,\ drop}$); $0.2 \times$ Max $V_{cc\ drop}$ is to provide a 14-dB safety margin.

†DC noise margin = difference between minimum V_{out} of driving gate and V_{in} required by driven gate to recognize a "1" or "0".

‡Peak instantaneous current that the driving device has to feed into each driven gate.

§The slow transition times and/or lower current of CMOS do not require one decoupling cap per chip.

SOURCE: ICT Inc.

TABLE 2.3 Signal Line Lengths for Various Logic Families above which Transmission Line Parameters Prevail

Logic family	Signal line length, mm (in)
Low-power Schottky (LS)	760 (30)
Schottky (S)	280 (11)
Advanced LS (ALS)	280 (11)
Advanced Schottky TTL	200 (8)
Advanced Schottky (AS)	150 (6)
Advanced CMOS technology	200 (8)
Emitter-coupled logic (ECL)	150 (6)

SOURCE: Technology Seminars Inc.

propagating pulse, one that is conceptually more straight forward for digital systems, compares the rise time to the path length without transforming into the frequency domain. The premise is to determine, for the transmission of a pulse of a given rise time, how long a conductor may be before a significant voltage difference is realized along its length. Conductors longer than this critical value are then regarded as transmission lines. However, irrespective of the precise definition that is used, the point at which a circuit becomes a transmission line is not defined by a single variable, but by the interplay of device rise time, conductor path length, and the relative dielectric constant (permittivity) of the medium.

There are three basic problems: first, how to lay out an interconnection system which has the proper dimensions and material properties to obtain a targeted impedance value; second, how to achieve a matched set of values from interconnections which have taken different routing, which implies different electrical values; and third, assuming that because of uncertainties in the properties of materials, the design rules, or processing, the impedance values are close but not within tolerance. The question remains as to how "trimming" can be performed to slew into the desired values. The classical representation of transmission lines for routing lines are in the forms of microstrip and stripline, shown in Fig. 2.8.

Unfortunately, with the advent of high-density interconnections, most systems will utilize multilayer printed circuit board and random X-Y gridding. If the outer layers were confined to lands and the signal layers of ground planes, this would approximate a stripline configuration. If signal lines were routed on the outer layers of the printed circuit board, this would approximate a microstrip if the immediate lower layer is a ground plane.

Signal layers stacked one upon another with the power and ground routed within each layer would constitute a planar configuration. For most complex multilayer designs, there will be a mixture of these con-

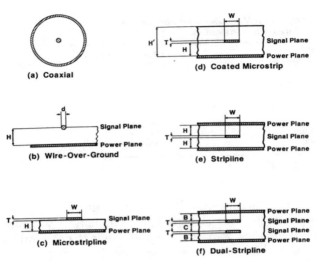

Figure 2.8 Typical transmission line configurations. *(From IPC-D-317.[1])*

figurations even within a single conductor line, which would make the task of predicting the exact impedance more difficult. Also, with high-speed devices it is important to match the transmission line imped-ance to the specific device input impedance. Device input impedances range from 50-Ω [GW] for ECL logic upward to 400 Ω for CMOS logic. Transmission line impedances of 50 Ω are relatively easy to manufac-ture. The higher-impedance values require decreased line widths along with lower-dielectric-constant materials and increased dielec-tric thickness. Additionally, the microstrip transmission line configu-ration could be favored because of its higher characteristic impedance.

Failure to match the transmission line impedance to the impedance at the end of the line will result in part of the signal being reflected back to the source, possibly causing false triggering of the device. Re-flections result in line ringing, which is a serious problem in packag-ing high-speed circuitry.

Line ringing must be controlled to avoid false triggering. If over-shooting can be controlled to less than 35 percent of the signal swing, then undershooting will be within acceptable limits. The actual value of impedance match must be maintained within a narrow range, prob-ably with 15 percent of the selected value. Line ringing is affected by device rise time, line length, termination and line impedance match, transmission line discontinuities, line loading, and termination type.

Transmission line discontinuities are affected by package type, cir-cuitry layout, and connectors. Discontinuities are caused by anything that creates a point of physical change in the signal path. Care must be taken to give priority to the high-speed-signal layout paths to minimize discontinuities which will in turn reduce the line ringing.

2.4.1.1 Velocity of propagation[6,8]. The two parameters of transmission line conductors of most importance in designing high-speed circuitry are velocity of propagation and characteristic impedance. The velocity of propagation is influenced by the effective dielectric of the transmission line configuration being used. In the case of the microstrip, both the insulating substrate material and the air surrounding the conductor must be considered, while with the stripline configuration only the insulating substrate material's dielectric constant is used.

The highest velocity of propagation is obtained by a conductor surrounded by air or a vacuum. In this case, the velocity of propagation is approximately 300 mm/ns (12-in/ns). Table 2.4 shows the velocity of propagation for various materials frequently used in electronic systems.

2.4.1.2 Characteristic impedance[7,9]. The second parameter of concern in transmission-line design is characteristic impedance Z_0, which is the single most important electrical parameter in determining the performance of high-speed designs. Each logic device family has a specific input impedance. With high-speed devices, it is important to have the characteristic impedance of the transmission line match this value of impedance. Any mismatch of impedance will result in problems with signal reflections, distortions, and crosstalk.

The characteristic impedance of a line is determined by the physical parameters of the line, such as capacitance and inductance. The equations that are used show a somewhat complicated relationship among conductor width w, conductor thickness h, the spacing between coplanar conductors, and the dielectric constant E_r and thickness of the insulating material. However, assuming that the resistive losses are very small compared to the reactive impedance and that the leakage current through the dielectric is also very small, the characteristic impedance of the interconnecting network is proportional to $(L/C)^{1/2}$. The values for L (self-inductance) and C (capacitance) in printed boards

TABLE 2.4 Electrical Properties of Various Materials

Substrate material	Dielectric constant	Velocity of propagation, mm/ns (in/nsc)
Air	1.0	300 (12.0)
Teflon glass	2.2	210 (8.1)
Epoxy glass	4.7	140 (5.5)
Polyimide glass	4.7	140 (5.5)
96% alumina	10.0	97 (3.8)
Silicon	11.7	89 (3.5)

are determined by the materials used and by the geometries of the conductors.

A range of characteristic impedance values, or "design window" exists for any system and is a function of both electrical and mechanical requirements. This window represents the optimum compromise between noise, delay, crosstalk, and producibility constraints. Once system characteristic impedance has been determined, design rules can be generated that not only ensure electrical performance but also ensure producibility.

The characteristic impedance of the printed circuit board interacts with the impedance of the system components to cause load and driver delays. The load-point delay is proportional to the product of characteristic impedance and the load capacitance. In addition, the printed circuit point characteristic impedance interacts with the driver output impedance as a voltage divider. The resultant lower driving voltage must charge up the line capacitance.

These delay factors, which react in opposite directions with characteristic impedance, can contribute to false switching in critically timed or high-speed systems unless the printed circuit board characteristic impedance is optimized to both the load capacitance and the driver output impedance. Noise enters the net from other nearby circuits (crosstalk) or from element switching. This crosstalk, combined with switching noise and reduced noise tolerance, can cause data errors. Like propagation delays, crosstalk and switching noise and noise tolerance interact with characteristic impedance in opposite ways. Figure 2.9[10] shows how total delay, total noise, and noise tolerance values will overlap to form a design window in which characteristic impedance values will achieve optimum system performance. The window can be derived for any logic net.

Basic physics dictates that maintaining a constant characteristic

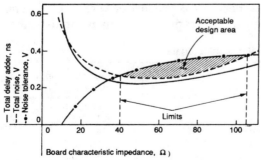

Figure 2.9 Characteristic impedance design "window." *(Courtesy of IBM Corp.)*

impedance Z_0 along a signal path provides maximum energy transfer with minimal attenuation or loss of waveform integrity. Toward this end, controlled-impedance multilayer printed circuit boards are used increasingly in demanding high-speed applications. Nonetheless, a logic signal traveling down a typical signal path still runs into such impedance discontinuities as IC pins, vias, connectors, signal branches, stubs, gate input capacitances, terminations, and etch-width variations. All these discontinuities degrade signal integrity to an extent. However, in general, the critical parameters in controlling characteristic impedance are conductor width, dielectric thickness, and dielectric constant of the insulating substrate material. Line thickness has a lesser effect.

2.4.1.3 Microstrip configuration[9]. The microstrip configuration, Fig. 2.10, has the printed circuit board conductors separated from one reference plane, either ground or power, by a dielectric that is usually glass-reinforced epoxy. (The equations shown for the various microstrip parameters assume that the major conductor dimension is much smaller than the dielectric thickness.) The key to the usefulness of the equations is the proper choice of the effective dielectric constant E_r.

Figure 2.11 shows that a change in dielectric constant changes the slope of the characteristic impedance Z_0 versus dielectric thickness h line, and Fig. 2.12 shows that for one dielectric constant E_r the variations of line widths w conductor thicknesses t and dielectric thicknesses h will result in parallel plots. Figure 2.11 leads to a fairly direct method of measurement of the effective dielectric constant for a microstrip configuration. Actual measurements of conductors using time domain reflectometry (TDR) will produce data that will lead to a plot much like Fig. 2.11. This is the most reasonable approach for printed circuit board manufacturers since the sample microstrips can be readily made.

The same method will allow a plot similar to Fig. 2.12, thus graphically defining the individual manufacturer's impedance control pa-

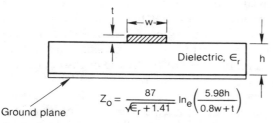

$$Z_0 = \frac{87}{\sqrt{E_r + 1.41}} \ln_e\left(\frac{5.98h}{0.8w + t}\right)$$

Figure 2.10 Microstrip transmission line configuration. *(From IPC-D-317[1])*

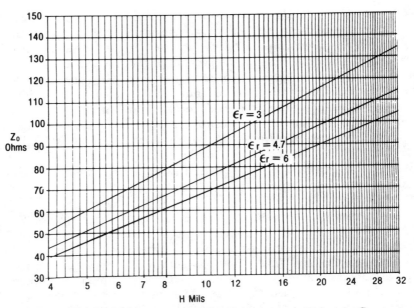

Figure 2.11 Microstrip Characteristic Impedance versus Dielectric Constant, 0.007-in-wide, 0.0014-in-thick conductor. *(Courtesy of Elco Corp.)*

rameters. Figure 2.13 is a plot with varying copper thickness t. Coated (imbedded) microstrip has the same conductor geometry as the uncoated microstrip. However, the effective relative dielectric constant E_r is different because the conductor is fully enclosed by the dielectric material. The equations for coated microstrip lines are the same for (uncoated) microstrip, with a modified effective dielectric constant.

If the dielectric thickness above the conductor is several mils or more, then the effective dielectric constant can be determined. For very thin dielectric coatings, the effective dielectric constant will be between that for uncoated circuits and that for a thickly coated line. For an embedded microstrip the dielectric constant along the edge of a conductor may be 35 percent lower than that between conductor and reference plane. This is due to the lack of glass fill next to the conductor edges. Also, if it becomes popular to leave photoresist on the conductor, there will be unfilled resin adjacent to the top of the conductor as well. The effective dielectric constant in this case will be strictly associated with one board manufacturer or another. Surface microstrip conductors will be in air or buried to some extent by solder mask. Thus prediction of surface microstrip impedance will depend on solder mask type and application process.

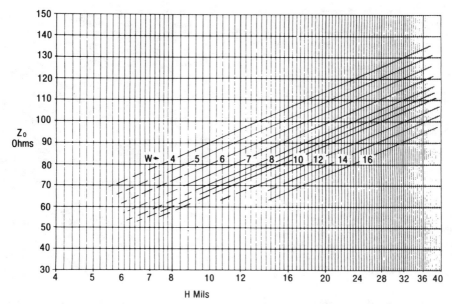

Figure 2.12   Microstrip characteristic impedance versus conductor width, dielectric constant of 4.7 and 0.0014-in-thick conductor.   *(Courtesy of Elco Corp.)*

Conductors are rarely rectangular in cross section, so the basic geometric assumptions are somewhat faulty. Conductors are actually various shapes from hourglass to trapezoid, depending upon etch process characteristics. Thus an agreement must be reached between board manufacturer and board designer as to how the line width is to be measured. Also, the dielectric composite material between conductor and reference plane is critical. If B-stage material is to be laminated in this layer, then the glass/resin ratio will be significantly different from a C-stage dielectric layer. Laminate supplier and individual board manufacturer processes govern this dielectric variable.

2.4.1.4 Stripline configuration[9,10]. The stripline configuration, Fig. 2.14, has reference planes above and below the conductor. Its electrical characteristics are a function of the same parameters as in the microstrip example. However, the relationships are different and result in lower values for characteristic impedance given in the same spacings. Since a typical multilayer printed circuit board of more than two signal layers has both stripline and microstrip geometries, the characteristic impedance will be different within the structure depending on the situation.

Neglecting any fringe edge capacitance, the stripline configuration

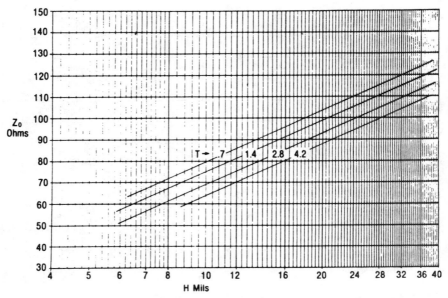

Figure 2.13 Microstrip characteristic impedance versus conductor thickness, dielectric constant of 4.7 and 0.007-in conductor width. *(Courtesy of Elco Corp.)*

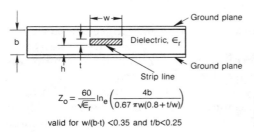

$$Z_0 = \frac{60}{\sqrt{\epsilon_r}} \ln_e \left(\frac{4b}{0.67\, \pi w(0.8 + t/w)} \right)$$

valid for w/(b-t) <0.35 and t/b<0.25

Figure 2.14 Stripline transmission line configuration. *(From IPC-D-317.[1])*

has twice the capacitance of the microstrip configuration, resulting in a characteristic impedance of half that of the microstrip configuration if t, w, and h are the same for both configurations (Fig. 2.15). The stripline impedance is always less than the microstrip impedance would be if one reference plane were removed. One approach suggests that the calculated upper and lower impedances may be averaged to get the stripline approximation. However, this will not work because it ignores the large capacitive coupling of the two planes.

Linewidth definition makes a significant difference, more so than with microstrip. The trapezoidal cross section must be taken into account, and as a result the etch characteristics must be predicted. Striplines which are not centered between the reference planes are

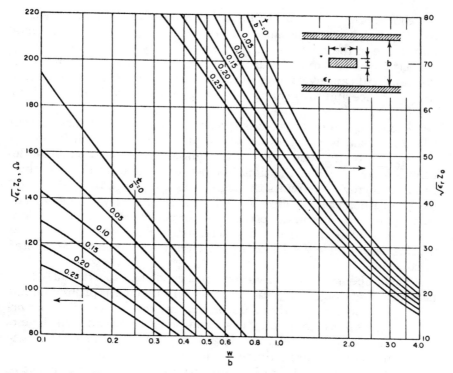

Figure 2.15 Stripline characteristic impedance versus conductor width and ground plane spacing. *(Courtesy of Elco Corp.)*

common. Asymmetrical characteristic impedance cannot be characterized as a straight line on a semilog plot.

2.4.1.5 Microwave printed circuit boards[11]. Microwave circuits are different than lower-frequency circuits because microwave frequencies are higher and therefore the wavelengths are shorter. Component lead lengths and body dimensions are a much larger percentage of a wavelength. Therefore they must be treated as transmission lines. The reactances of parasitic lead inductances and shunt capacitances become significant and cannot be ignored at microwave frequencies.

The most important parameters of a microwave printed circuit board are the dielectric constant and dissipation factor of the dielectric, the circuit conductor dimensions, and the printed circuit board thickness. All of these parameters affect the performance of the microwave circuit. The dielectric constant of microwave printed circuit boards must be carefully controlled. This is in contrast with printed circuit board materials used for lower-frequency circuits. The dielectric constant of microwave printed circuit boards must be kept within reasonable tolerances because the electrical length and characteristic impedance of the microwave transmission lines are a function of the dielectric constant of the board. Therefore, variation in the dielectric constant will cause variation in the electrical length and characteristic impedance and thus changes in the microwave performance of the circuit.

The dissipation factor, or loss tangent, of microwave printed circuit board material is much lower than that of materials used for lower-frequency circuits, such as epoxy fiberglass. The dissipation factor determines how much of the signal is lost (dissipated) in the dielectric. In most microwave circuits the signal energy passes through the dielectric rather than in discrete components as in low-frequency circuits. The conductor on the printed circuit board is the circuit in a microwave printed circuit board. Therefore, the dissipation factor has much more effect upon losses in the circuit on a microwave printed circuit board.

It should be noted that environmental conditions, particularly temperature, may have a significant effect on microwave printed circuit board parameters and therefore on the microwave circuit's performance. The environmental effects should be accounted for in the circuit design. Unfortunately, the data regarding environmental effects on printed circuit board parameters is often not available.

Variations in printed circuit board parameters can cause microwave printed circuit boards to not meet their performance specifications. The circuit designer has four alternatives to deal with this problem:

1. *Tighten the tolerances on printed circuit board parameters:* This alternative makes things easy for the printed circuit board designer but difficult for the material manufacturer and printed circuit board fabricator. Usually this is not a realistic alternative since the tolerances on printed circuit board parameters are fixed and the circuit designer must live with them. However, in some situations the tolerances may be tightened. This usually results in higher costs for the printed circuit board, although the total product cost may or may not be higher.

2. *Design the circuit to meet specifications with any combination of parameter variations:* This alternative is available for some circuits. However, designing to accommodate all parameter variations may result in an increase in size. There will probably also be an increase in nonrecurring engineering time since the designer may have to do several designs before arriving at the one which will meet specifications with all parameter variations.

3. *Allow some percentage of printed circuit boards to be rejected be-*

cause of inadequate performance: This alternative assumes correctly that the variation in printed circuit board parameters has some statistical properties. Therefore there is a low probability that all the parameters will go in the same direction, as in the example on all the printed circuit boards produced. Armed with the statistics of the parameter variation, the designer can calculate what percentage of printed circuit boards will not meet specification. It may then be economically feasible to discard the printed circuit boards that do not meet specification. This assumes that it will be possible to test the circuit individually after fabrication and that the statistics of parameter variation are available. Often, neither is the case.

4. *Tune the circuit after fabrication to meet specification:* In choosing this alternative the designer must trade off fabrication cost versus the labor cost for tuning. Tuning is naturally more attractive when only a few printed circuit boards are being produced. However, as the number of printed circuit boards increase it becomes more economically feasible to put more effort into the nonrecurring design and reduce or eliminate tuning. Basically, the choice of how to deal with microwave printed circuit board parameter variation is an economic one. One approach will be more cost-effective than the other depending on the performance requirements of the circuit and on the quantity being produced. The nonrecurring engineering material cost, fabrication cost, and recurring labor cost must all be considered.

2.4.1.6 General design guidelines[11]. The following guidelines provide a practical approach to transmission line layout:

- *Review the CAD layout as it progresses:* Changes made after the layout is complete will have a tremendous impact on nearby signals and could even increase the layer count.

- *Make critical signals first priority in the layout, so they can be handled most effectively:* Document a list of critical signals, such as clocks and strobes, along with layout rules to route first. Review and approve their pen plots before general layout begins.

- *Keep conductors short whenever possible:* Lumped nets avoid many problems associated with transmission lines. Most impedance discontinuities can be ignored. "Short" is a relative term and is related to rise time.

- *Use the slowest logic family that can meet timing requirements:* Slower logic allows longer lines to still appear lumped, easing layout constraints. Lower EMI emission is an additional benefit.

- *Keep loading along distributed nets equal and well balanced:* This aspect of layout tremendously affects signal integrity. When loads are distributed evenly on a transmission line, their input capacitance and stub capacitance combine to increase the distributed capacitance. This approximates a lowering of the line's effective impedance. When the loading is well balanced, the lowered effective impedance is relatively constant, so reflections are minimized. When loading is unbalanced, however, the various effective impedances of the different sections create multiple reflections. Unfortunately, this is very common in CAD layouts. If routability restrictions make changes in the layout impossible, reflections still can be reduced significantly by varying the etch width of different sections to change the characteristic impedance Z_o so that the various loading densities create matched effective impedances.

- *Avoid branching in the layout of critical signals:* This is a very common pitfall of CAD layout systems. If the signal branches, the impedance at the branch is $Z_o/2$. This discontinuity creates significant reflections. If routability restrictions require branching, it is important to try to keep each branch short enough to meet the lumped criteria of the logic family.

- *Keep transmission-line stubs less than 76 mm (3 in) for TTL, CMOS, ECL-10K, and less than 25 mm (1 in) for ECL-100K:* Because stubs are essentially branches, it is important to keep them clearly lumped. On backplane–daughter-board systems, this means keeping all drivers and receivers as close as possible to the edge-board connector.

- *Proper grounding practices are mandatory for signal integrity:* If proper return current paths are not provided, even the best layout will not perform satisfactorily.

- *When using logic devices with rise times of 1 ns or less, use 45° angles, as opposed to right angles, in the conductor path:* Also, keep vias to a minimum. Right angles and vias create impedance discontinuities. With very fast edge rates, even these small reflections can affect signal integrity significantly. (Laying out within these constraints can limit routability severely, which is another reason to use slower devices if timing allows.)

- *Avoid long test-point etches:* Attaching a long unterminated etch path for a test point creates numerous reflections. Keep test-point etches very short or use a separate buffer if possible.

- *Minimize the use of vias:* Vias are yet another impedance source that can degrade high-frequency circuit performance. By eliminating vias, either by using a minimization program or by having an experienced designer fine-tune the circuit layout, it is possible to eliminate reflection sites. Historically, designers have tried to minimize vias because vias lower manufacturing yields and increase production costs. In addition, designs that contain few vias are easier to route than designs with many vias, and reducing the number of vias also simplifies the implementation of design changes.

2.4.2 Crosstalk[4,5]

To complicate printed circuit board design, there is the problem of signal isolation, or crosstalk. The need for high density has lead to finer conductor lines and closer spacing. With the closeness of the conductors, and higher signal speeds, the coupling of signals into adjacent conductor lines becomes greater and introduces noise and false signals into systems.

Crosstalk is a problem at high frequencies because, as operating frequencies increase, signal wavelength becomes comparable to the length of some of the interconnections on the printed circuit board. Under these conditions, interconnections actually become antennas and begin broadcasting. Three types of signal coupling determine the amount of crosstalk in a circuit: inductive coupling, capacitive coupling, and radiative coupling.

Inductive, radiative, and capacitive coupling all decrease with increasing distance between source and receiver; most crosstalk can be attributed to adjacent wires. To minimize crosstalk, then, first examine a circuit's interactions with its nearest neighbors. Because parallel and adjacent wires on a printed circuit board layer interact both radiatively and inductively, minimize the distance over which adjacent wires are parallel.

The printed circuit board interconnection itself adds to the propagation delay of every signal the wire carries. Not only do interconnections decrease the operating speed of circuits, these lines also distort output signals. The contribution of an interconnection to a circuit's delay can exceed a high-speed device's inherent gate delay. For example, the propagation delay of an electrical signal in copper on G-10 glass epoxy is approximately 2 ns/ft, so a 100-mm (4-in) line adds about as much signal delay to a design as a 100K ECL gate does. Therefore, circuit simulations must take into account the length of interconnections.

When a signal path is coupled to an adjacent path that would affect a circuit at the receiving end of the line, it is called forward crosstalk. When a signal is coupled to a signal at the sending end, it is called backward crosstalk. Forward crosstalk is normally much smaller than the backward crosstalk. For minimum crosstalk, the design should minimize close spacing of lines and low-impedance lines should be used. The lower the characteristic impedance, the less crosstalk. The ideal situation is to make the source impedances low to minimize interference due to capacitive coupling and load impedances high to reduce the amount of voltage induced in other circuits due to inductance coupling.

2.4.2.1 Inductive coupling[12]. Impedance is the culprit in most ground layout problems, and inductive pickup is one cause of crosstalk. Because the currents that flow in digital circuits tend to be spiked, the inductance of the path, not trace resistance, dominates impedance. Conductor resistance and the width of the conductor has little effect on impedance.

Unlike resistance, inductance does not depend on the length or cross section of the conductors. Rather, inductance depends on the geometry of the circuit. On the printed circuit board with the single ground path, the current loop is quite large (Fig. 2.16). The source and destination chips are quite close together, but the ground layout forces maximum inductance.

A useful circuit has many loops in which currents can circulate, but each current individually traces out its own loop (Fig. 2.17). The geometry of each loop determines its inductance in the same manner as in the single-turn circuit. The current loop for a given signal can be identified by tracing the signal from source to destination. The current follows the path of least impedance, that is to say, it follows the path of minimum loop area. The layout of power and ground conductors is critical to the noise properties of the circuit because they provide the signal return paths. In addition, loops in the ground layout of

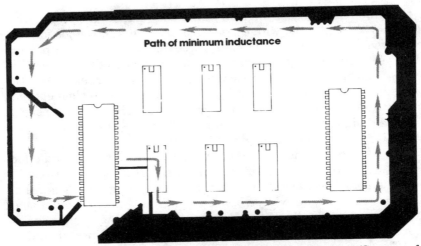

Figure 2.16 Printed circuit board layout with single ground path. *(Courtesy of Intel Corp.)*

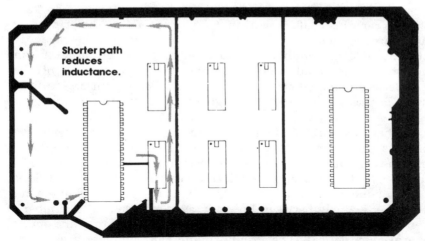

Figure 2.17   Printed circuit board layout with multiple ground paths.   (*Courtesy of Intel Corp.*)

a printed circuit board are not "ground loops." Their presence does not increase noise but rather decreases it.

In a printed circuit board layout, what has generally come to be called a "ground loop" problem is really a problem of common-ground impedance. Noise is caused by digital current spikes being forced to share a return path with low-level analog signals. Thus, when digital and low-level analog signals are present on the same printed circuit board, a separate return path for the analog signals should be provided.

2.4.2.2   Capacitive coupling.   The capacitive-coupling effect links any two wires that are at equal voltage levels (equipotential), because the wires form a capacitor and couple into each other's signals through the nonconducting medium that separates the wires. The electric field between two capacitor surfaces (and thus the intensity of capacitive coupling) varies in proportion to the permittivity of the nonconducting medium and to the parallel component of the areas of the conductors. Capacitive coupling decreases with increasing distance between capacitor surfaces.

To minimize the crosstalk caused by capacitive coupling, high-frequency designs should include ground planes under each signal layer. Ground planes limit to small sections the crosstalk caused by capacitive coupling between adjacent layers (at equipotentials). Because the width and thickness of signal lines and their distance from ground is constant, the capacitive-coupling contribution to impedance remains uniform within ± 5 percent across the printed circuit board. Although a fixed impedance does not reduce capacitive coupling, it does simplify the modeling of propagation delays and coupling effects. In addition, capacitive coupling can cause interference between printed circuit board layers. To prevent crosstalk between layers, route wires on neighboring layers in an orthogonal fashion.

Another way that a designer can reduce capacitance of back-to-back conductors on different layers is to reduce the width of the conductor, since the capacitance is inversely proportional to the conductor width. However, reducing conductor width increases the series impedance of the conductor and a trade-off must be made. A ground plane may be used effectively between layers in systems where the number of interconnections sensitive to crosstalk is too large to make spatial separation practical. The ground plane can completely eliminate the capacitive coupling between signal conductors on separate layers.

However, the effectiveness of the ground plane is limited because the impedance of ground wires and the conducting layer itself make it impossible to maintain all parts of the plane at signal ground potential. Thus, the need for low-impedance return to the ground plane cannot be overstressed.

Some general and short rules that a printed circuit board designer can use to reduce noise generation and susceptibility of circuits include:

- Do not put too much emphasis on inductance per unit length figures. Inductance is not a property of the conductors in the circuit. It is a property of the geometry of the complete circuit.

- Remember that the term "ground loop" has nothing to do with loops in the ground layout. Loops in the ground layout are good, not bad.

- Lay out the power and ground distribution conductors first, not last. They do not just supply dc power, they are also the return path for all the signal currents on the printed circuit board. An interconnected grid of narrow conductors is better than a few wide conductors.

- For effective decoupling, mind the layout of the loop formed by the integrated circuit and its decoupler capacitor. The area of this loop is more important than the value of the capacitor.

- Supply and ground conductors for analog circuits should be kept separate from supply and ground conductors for digital circuits on the same printed circuit board, so that they do not become part of the multiplicity of digital return paths.

2.4.3 Electromagnetic interference[13–15]

The concern for electromagnetic compatibility (EMC) in electronic systems has risen since the Federal Communications Commission (FCC) proclaimed that there shall be no more pollution of the electromagnetic spectrum. Still, designers have not yet fully come to grips with a major source and victim of electromagnetic interference (EMI), the printed circuit board. Whereas cables, card racks, and enclosures can be built to contain or keep out EMI by shielding, the printed circuit board is often the system component that generates the offensive energy or succumbs to it. Thus the most critical stage for addressing EMI is during printed circuit board design.

Printed circuit boards containing high-frequency digital logic circuitry may excite small resonant loop antennas, run patterns on the printed circuit board, and exceed the radiated emissions limits of FCC, VDE, and military standards. Run-to-run coupling on some printed circuit boards causes logic errors and other circuit noise problems. These problems are usually associated with poor layout, poor grounding, and inadequate power run decoupling for high-speed circuits.

2.4.3.1 Logic selection.
Logic selection can ultimately dictate how much attention must be given to EMC in the circuit design. The first guideline is to use the slowest-speed logic that will do the job. Logic speed refers to rise and fall times and gate response. Many emissions and susceptibility problems can be minimized if a slow-speed logic is used.

It is also not unlikely to see printed circuit boards that are susceptible to incoming noise signals with frequencies way above the operating clock frequency of the circuit. This coupling phenomenon is a function of the logic frequency bandwidth, and often is referred to as audio rectification. Refer back to Table 2.2 for the characteristics of popular logic families. The rise and fall times, as well as the associated logic bandwidths are also given.

The type of logic to be used is normally an early design decision, so that control of edge speeds and, hence, emissions and susceptibility are available early. Of course, other factors such as required system performance, speed, and timing considerations must enter into this decision. The use of slow-speed logic, however, does not guarantee that EMC will exist when the circuit is built; so proper EMC techniques should still be implemented consistently during the remainder of the circuit design.

2.4.3.2 Circuit placement. Schematics tell little or nothing about how systems will perform once the printed circuit board is fabricated, assembled, and powered. A circuit schematic is useful to the design engineer, but an experienced EMC engineer refers to the printed circuit board when troubleshooting. By controlling the printed circuit board layout in the design stage, the designer realizes two benefits:

- A decrease in EMI problems when the circuit or system is sent for EMI or quality assurance testing
- The number of EMI coupling paths is reduced, saving troubleshooting time and effort later on

Printed circuit boards are generally laid out so that the higher-speed logic is located near the edge connector and the lower-speed logic and memory, if applicable, are located furthest from the connector. This tends to balance common-impedance coupling, radiation, and crosstalk.

If the highest logic speed is below TTL, then layout is relatively unimportant from an EMI point of view. The single exception is when optical isolaters, isolation transformers, or filters are used. In this case these devices should be located as close to the edge-board connector as possible.

Isolation of the I/O from digital circuitry is important where emissions or susceptibility is a problem. For the case of emissions, a frequently encountered coupling path involves digital energy coupling through I/O circuitry and signal traces onto I/O cables and wires, where the latter subsequently radiate. When susceptibility is a problem, it is common for the EMI energy to couple from I/O circuits onto sensitive digital lines, even though the I/O lines may be "optocoupled" or otherwise supposedly isolated. In both situations, the solution often lies in the proper electrical and physical isolation of analog and low-speed digital lines from high-speed circuits. When high-speed signals are designed to leave the board, the reduction of EMI is usually performed via shielding of I/O cables.

High-speed logic components should be grouped together. Digital interface circuitry and I/O circuitry should be physically isolated from each other and routed on separate connectors, if possible. Oscillators and clock circuits should be located near the center of the printed circuit board. This location minimizes coupling to I/O runs and places the higher-frequency sources in the center of the power and ground distribution system. High-frequency switching currents are then supplied uniformly, and distributed through a 360° source pattern, rather than exciting the edge of a printed circuit board with a hot spot.

Memory circuits should be located away from I/O circuits and their runs, preferably near the center of the printed circuit board. These circuits draw high current transients during switching and excite the printed circuit board's ground system. A central location on the printed circuit board allows these currents to be supplied uniformly by the printed circuit board's natural and discrete component capacitance.

2.4.3.3 Power distribution. Isolated digital and analog power supplies should be used when mixing digital and analog circuitry on a printed circuit board. Often, the power supply design determines whether the system can or cannot run by itself; one problem involves common impedance coupling between digital and analog circuitry. This impedance is "shared" by two circuits and may be a common ground impedance or a common power source impedance. The design preferably should provide for separate power supply distribution for both the analog and digital circuitry.

Single-point common grounding of the analog and digital power supplies should be performed at one point and one point only—usually at the motherboard power supply input for multicard designs, or at the power supply input edge-board connector on a single-card system. The main feature of good power bussing is low impedance and good decoupling over a large range of frequencies. Achieving a low-impedance distribution system requires taking into account the following considerations.

Controlled-impedance transmission lines. The bandwidth of many common logic families extends from 10 to 100 + MHz as shown in Table 2.4. This spread means that the circuit must be capable of conducting RF energy in the HF and VHF bands. Although it may be painful for the designer to think in terms of transmission lines and characteristic impedance, these are the phenomena that are encountered in high-speed (high-frequency) logic design.

The impedance of an isolated power supply and return bus increases with frequency to become self-limiting for high-speed logic. Thus, what is actually needed is a lower impedance or impedance control. Table 2.5 shows the characteristic impedance of three different transmission lines as a function of their geometry. Thus, power distribution for serving logic should be pictured as distributed over a transmission line. Any one of the three configurations may be viewed as a possible

TABLE 2.5   Characteristic Impedance of Different Conductor Pairs

	#1	#2	#3
	Z_{01} (parallel strips, w, t, h)	Z_{02} (strip over gnd plane, w, h)	Z_{03} (strips side by side, t, D, w)
W/h or D/W	Parallel Strips*	Strip Over Gnd Plane*	Strips Side by Side†
0.5	377	377	NA
0.6	281	281	NA
0.7	241	241	NA
0.8	211	211	NA
0.9	187	187	NA
1.0	169	169	0
1.1	153	153	25
1.2	140	140	34
1.5	112	112	53
1.7	99	99	62
2.0	84	84	73
2.5	67	67	87
3.0	56	56	98
3.5	48	48	107
4.0	42	42	114
5.0	34	34	127
6.0	28	28	137
7.0	24	24	146
8.0	21	21	153
9.0	19	19	160
10.0	17	17	166
12.0	14	14	176
15.0	11.2	11.2	188
20.0	8.4	8.4	204
25.0	6.7	6.7	217
30.0	5.6	5.6	227
40.0	4.2	4.2	243
50.0	3.4	3.4	255
100.0	1.7	1.7	293

*Mylar dielectric assumed: $\epsilon_r = 5.0$
†Paper base phenolic or glass epoxy assumed: $\epsilon_r = 4.7$
$Z_{01} = (377/\sqrt{\epsilon_r})(h/W), \text{for } W > 3h \text{ and } h > 3t$
$Z_{02} = (377/\sqrt{\epsilon_r})(h/W), \text{for } W > 3h$
$Z_{03} = (120/\sqrt{\epsilon_r})\ln_e(D/W + D/^2 - 1). \text{for } W > > t$
$D > >$ nearby ground plane
SOURCE: Don White Consultants

method of routing power supply (or signal) conductors. The most important feature of Table 2.5 is the vast difference in impedance between the parallel strips and strip over ground plane compared with the side-by-side configuration.

TABLE 2.6   Ground Plan Impedance for 0.0014 In-Thick Copper Foil

Frequency	Impedance, mΩ/sq.	Frequency, MHz	Impedance, mΩ/sq.
10 Hz	0.812	10	1.53
100 Hz	0.813	20	1.89
1 kHz	0.817	30	2.20
10 kHz	0.830	50	2.71
100 kHz	0.871	70	3.15
1 MHz	1.01	100	3.72
2 MHz	1.10	200	5.22
3 MHz	1.17	300	6.39
5 MHz	1.29	500	8.25
7 MHz	1.39	700	9.76

SOURCE: Don White Consultants.

As a natural extension to the evolution of reducing the power supply and return impedance, a complete ground plane is imagined to have the lowest impedance. To this end, Table 2.6 represents the impedance of 1-oz copper foil. In comparing the impedances within Table 2.6, it is noted that at low frequency, little change exists with a change in frequency. This is because the impedance is essentially resistive. Conversely, for high frequency, such as above 10 MHz, the impedance is proportional to the square root of frequency due to skin depth effects.

In comparing the values in Table 2.6 with conductor impedances in Table 2.7, it is noted that the former corresponds to impedances ranging from two to four orders of magnitude less than the latter. Thus, a two-dimensional planar surface is highly desirable for distributing power and for use as a power or signal return. Therefore, except for rather severe conditions, the impedance drop in the ground plane would be small compared to most analog sensitivities and all logic noise immunity levels.

Multilayer printed circuit boards.   Power and ground planes offer the least overall impedance. The use of these planes leads the designer closer to a multilayer printed circuit board. At the least, it is recommended that all open areas on the printed circuit board be "landfilled" with a 0-V reference plane so that ground impedance is minimized. Multilayer printed circuit boards offer a considerable reduction in power supply impedance, as well as other benefits. From Table 2.5 the impedance of a multilayer power or ground plane bus is very small, on the order of an ohm or less, assuming a w/h ratio greater than 100. For high-density, high-speed logic applications, the use of a multilayer printed circuit board is almost mandatory, and printed circuit boards with 10 layers or more, including signal planes and power supply planes, are common in complex systems.

TABLE 2.7   Characteristic Impedance for 1-mm (0.040-in) Wide by 0.04-mm (0.0014-in) Thick Conductor

Frequency	Conductor length, mm (in)			
	1 (0.04)	3 (0.12)	10 (0.4)	30 (1.2)
10 Hz	5.74 mΩ	17.2 mΩ	57.4 mΩ	172 mΩ
100 Hz	5.74 mΩ	17.2 mΩ	57.4 mΩ	172 mΩ
1 kHz	5.74 mΩ	17.2 mΩ	57.4 mΩ	172 mΩ
10 kHz	5.76 mΩ	17.3 mΩ	57.9 mΩ	174 mΩ
100 kHz	7.21 mΩ	24.3 mΩ	92.5 mΩ	311 mΩ
1 MHz	44.0 mΩ	173 mΩ	727 mΩ	2.59 Ω
2 MHz	87.5 mΩ	344 mΩ	1.45 Ω	5.18 Ω
3 MHz	131 mΩ	516 mΩ	2.17 Ω	7.76 Ω
5 MHz	218 mΩ	861 mΩ	3.62 Ω	12.9 Ω
7 MHz	305 mΩ	1.20 Ω	5.07 Ω	18.1 Ω
10 MHz	437 mΩ	1.72 Ω	7.25 Ω	25.8 Ω
20 MHz	874 mΩ	3.44 Ω	14.5 Ω	51.7 Ω
30 MHz	1.31 Ω	5.16 Ω	21.7 Ω	77.6 Ω
50 MHz	2.18 Ω	8.61 Ω	36.2 Ω	129 Ω
70 MHz	3.05 Ω	12.0 Ω	50.7 Ω	181 Ω
100 MHz	4.37 Ω	17.2 Ω	72.5 Ω	258 Ω
200 MHz	8.74 Ω	34.4 Ω	145 Ω	517 Ω
300 MHz	13.1 Ω	51.6 Ω	217 Ω	776 Ω

SOURCE: Don White Consultants

Where both a supply and return plane are close together and form a transmission line, the impedance is no longer the low surface impedance of each. Table 2.8 is a tabulation for several values of h/d and ϵ_r. Basically, for h/d less than about 0.005 and greater than 3 (estimated to be most of the printed circuit boards in multilayer printed circuit

TABLE 2.8   Characteristic Impedance of Printed Circuit Board Supply-Return Planes, Ω

Dielectric constant	h/d ratios of parallel plane								
	0.0005	0.0007	0.001	0.002	0.003	0.005	0.007	0.010	0.020
1.0	0.19	0.26	0.38	0.75	1.13	1.88	2.64	3.77	7.54
2.0	0.13	0.19	0.27	0.53	0.80	1.33	1.86	2.67	5.33
Glasscloth*	0.12	0.17	0.24-	0.49	0.73	1.22	1.70	2.43	4.87
3.0	0.11	0.15	0.22	0.44	0.65	1.09	1.53	2.18	4.35
4.0	0.09	0.13	0.19	0.38	0.57	0.94	1.32	1.88	3.77
5.0	0.08	0.12	0.17	0.34	0.51	0.84	1.18	1.69	3.37
7.0	0.08	0.11	0.15	0.31	0.46	0.77	1.08	1.54	3.08
7.0	0.07	0.10	0.14	0.29	0.43	0.71	1.00	1.42	2.85
8.0	0.07	0.09	0.13	0.27	0.40	0.67	0.93	1.33	2.67
9.0	0.06	0.09	0.13	0.25	0.38	0.63	0.88	1.26	2.51
10.0	0.06	0.08	0.12	0.24	0.36	0.60	0.83	1.19	2.38

Glasscloth = woven or nonwoven Teflon, PTFE laminate, ϵ_r = 2.4
SOURCE: Don White Consultants.

boards), characteristic impedance is less than 1-Ω. Thus, the potential drop at the logic chip for 20 mA of switching current is less than 20 mV, or well below that of the noise-immunity level. Because this is a planar transmission line, the drop is not common-impedance-coupled to all of the other chips supplied by the same supply and return plane. Thus, power plane distribution is an excellent approach for controlling high-speed logic waveshape and performance.

Double-sided printed circuit boards. The problem with multilayer printed circuits boards is the increased cost of design and fabrication and increased difficulty in printed circuit board repair. Thus it is sometimes the goal to produce double-sided printed circuit board designs that behave similarly to multilayer boards. Double-sided printed circuit boards can and have been designed to accomplish the natural noise cancellation inherent with multilayer designs. For these applications, the following step-by-step process is recommended:

1. *Board I/O pins:* Select evenly spaced power and ground pins for signal I/O to the printed circuit board. For fast-rise-time signals, i.e., Schottky, it is recommended that ground pins be spaced no greater than 25 mm (1-in) apart along the I/O connector. For emitter-coupled logic (ECL) series and other faster signal interfaces, reduce this spacing to 13 mm (0.5-in). It is not advisable to depend upon a single ground pin at one end of the connector.

2. *Dedicated board ground runs:* The I/O ground pins are then connected to prededicated ground runs that flow onto the printed circuit board, evenly spaced. These ground runs project up through the planned rows of circuits and are temporarily placed on the printed circuit board prior to adding circuits; 0.5 mm (0.020 in) wide is sufficient, but their location is very important. They may be staggered somewhat to relieve congestion; however, this practice should be minimized if possible.

3. *Noisy and critical circuits installation:* Clock circuits and their output runs are placed on the printed circuit board first. Then locate the oscillator and buffer in the center. Locate clock divide-by circuits nearby. Clock runs should be added either to the backside (opposite) the prededicated ground run, or adjacent to it on the same side as the printed circuit board. If the clock run lies adjacent, it is recommended that a second ground run be added; thus, the clock run lies between two ground runs. Where clock runs fan out and drive long distances such as 200 to 250 mm (8 to 10 in), it is recommended that parallel resistive termination be designed in at the load end. Other circuits that should be located in the initial design layout include memory and I/O circuits to remote units and equipment.

4. *The board design may then continue according to normal methods:* The predicated ground runs may be moved somewhat and/or relocated to the other side of the printed circuit board, or many crossovers may be added, i.e., side-to-side. Circuit grounds are connected to the printed circuit board's predicated ground runs as continuously as possible. Cross jumper runs between the predicated ground runs are then added on either side of the printed circuit board. These should be added at each circuit where possible. Cross jumper run width is not critical; 0.25 mm (0.010 in) in width is sufficient.

5. *The printed circuit board's ground grid system is then subjected to landfilling on both sides:* This process involves widening the ground grid runs and filling in open voids with grounded diamond grid patterns. The extent of landfilling is important; the more ground, the smaller the loops.

When complete, the double-sided printed circuit board should have the following characteristics:

- Circuits are referenced together in both directions, vertically and horizontally.
- Overlay of the artwork indicates that the printed circuit board is almost opaque due to the ground landfilling.

Decoupling. This design feature is of paramount importance. The decoupling capacitor adjacent to a chip serves two functions, i.e.,

- It supplies transient current to the chip during gate switching.
- It limits the size of a potential radiating loop associated with the power supply to a switching gate.

The limitation to the frequency characteristics of any capacitor is the self-resonance caused by the inductance of the capacitor lead length. The ideal capacitor should have low loss and should remain capacitive over the applicable frequency range. Aluminum and tantalum electrolytics should be used with caution for high-frequency decoupling since they are self-resonant at a few hundred kilohertz and a few megahertz, respectively. Z5U ceramic capacitors resonate between 1 and 20 MHz, depending on the formulation and packaging, but are useful for decoupling when the EMI frequency is below 50 MHz.

The second function of decoupling capacitors is to reduce power supply radiating loops. Since the current and frequency usually are predetermined by the type of logic selected, it is necessary to minimize the area of the logic current loop to reduce radiation. Minimal area can be accomplished by proper printed circuit board layout.

2.4.4 Signal routing

Proper signal routing is an important step toward achieving EMC in the printed circuit boards. Many of the same guidelines concerning component layout are continued. Here, the key word is isolation—between high-voltage and current lines (such as relay drivers), I/O lines, and analog and digital conductors. A surprising degree of coupling that will often make the difference between success and failure of the EMC performance of a printed circuit board can occur between two adjacent conductors. Finally, impedance matching becomes important for very high speed logic to minimize signal distortion caused by reflections.

A common practice is to first route signal conductors on the printed circuit board and then fill in the power supply bus wherever there is room. It is usually more desirable, for EMC considerations, to construct the power distribution system first and route the signal conductors thereafter. High-frequency clock and clock-derived lines should be as short as possible to minimize radiating loops.

Because the O-V plane constitutes the return path for digital signals, the use of a landfilled O-V ground plane on the printed circuit board can offer reduction of loop areas associated with clock and high-frequency lines. If the ground plane is very extensive over the surface of the printed circuit board and a signal conductor is routed directly above it, the size of the loop is approximately the length of the conductor times the thickness of the dielectric of the printed circuit board, resulting in a relatively small loop.

When high-speed circuitry communicates with I/O circuitry, coupling through I/O circuits to wires and cables which go off-board must be minimized. To accomplish this, consider using dedicated signal/return lines (signal on one side of the printed circuit board, return on the other) between the high-speed circuitry and I/O circuitry. Using capactive decoupling on high-frequency lines will "soften" edges and reduce crosstalk; the size of the capacitor depends on how much distortion the desired signal can tolerate.

In addition, high-speed signal conductors should be routed away from I/O conductors which might otherwise capacitively couple energy to the I/O circuitry. This coupling path is quite common where emissions and susceptibility are a problem, often being the major contributor to an EMI situation.

The use of O-V "guard traces" between the high speed and the I/O lines will provide some shielding of the I/O from the high-frequency energy. Thus, when high-density printed circuit boards are designed, the use of guard traces should be examined. A landfilled O-V ground plane can perform the function of the guard trace, if it is filled in between high-speed and I/O lines.

2.4.3.5 Printed circuit board grounding. Many designs using high-frequency clocks (20 MHz and above) require the connecting of the printed circuit board's ground systems to the metal chassis of the unit. This common RF grounding concept reduces the impedance between the noisy printed circuit board and chassis, allowing the chassis to serve as a shield. The common connection also enables additional signal line filtering relative to the chassis at other I/O connectors within the equipment, if necessary.

Printed circuit boards that mount within this unit via stand-offs should provide jumper capabilities to tie each printed circuit board's ground grid to the unit via the stand-offs and mounting screws. This tying to the unit can be done either directly or with small capacitors which provide a good RF connection but which remain isolated at powerline frequencies.

Similar methods can be used with plug-in printed circuit boards, where chassis ground can be achieved through metallic card guides. The printed circuit board's edges are landfilled to interface with the card guides, and small capacitors are added to jumper the printed circuit board's ground grid to the plug-in ground on the edge of the printed circuit board.

2.4.4 Testability[3]

The following guidelines should be used on all logic designs to improve the testability of the printed circuit board assembly.

- Connect the set, clear, or reset of all flip-flops to reset terms. Power-up preset is ideal (Fig. 2-18).

- Connect the clear or master reset of counters to reset terms or provide preset logic (Fig. 2-19).

- Provide a method to disable free-running oscillators, and provide a pin-out to an unused I/O pin on the oscillator clock line.

- Provide outputs to unused I/O pins for feedback or "deep" logic loops. This also applies for serial logic.

- Keep direct output control, wherever possible, on all IC devices to enhance their testability.

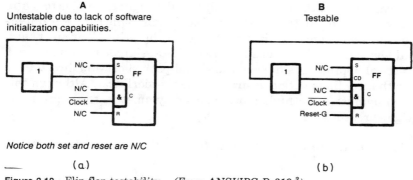

Figure 2.18   Flip-flop testability.   *(From ANSI/IPC-D-319.[3])*

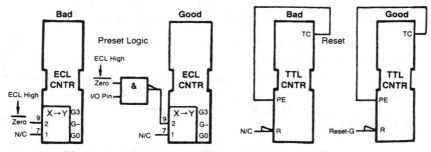

Figure 2.19   Master clear for counters.   *(From ANSI/IPC-D-319.[3])*

- Provide jumper lines from the delayline clock circuits to unused I/O pins, so that intermediate clocks may be controlled.

- Provide test points to appropriate conductors in untestable circuits.

2.5   Analog Circuit Design[15,16]

Analog printed circuit board design differs in some very fundamental ways from its digital counterpart. These differences are inherent in the divergent approaches to data structure that the two circuit techniques employ. A digital signal consists of quantified data, i.e., numbers, or "digits" represented by a stream of logic levels. Such a signal must be relatively noise-free, and its edges must be sharp, but its circuitry can tolerate a large amount of distortion without compromising the accuracy of the data. By contrast, the waveform of an analog signal is an exact replica of the data it represents. It is "analogous" to that data, and the data can be no more accurate than the integrity of the signal representing it.

The challenge to the printed circuit board designer, then, is to maintain the absolute purity of these analog signals. This can be a tall order when components are jam-packed onto the printed circuit board, signal levels are low and sensitive, and digital signals are intermixed. A practical approach, is to keep the signal path short and direct, even where that conflicts with a desire for visual orderliness. Although this may go against some higher principle of symmetry, there are some very good reasons for doing it this way:

- Related parts are kept very close together, and connections can be very short indeed.

- Circuits are separated, and unwanted interaction among them is kept to a minimum.

- The surrounding space can be flooded with ground, improving the ground-plant distribution and further isolating the circuits.

- Changes can be made to one area with minimum impact on the rest of the printed circuit board.

- Over time, a library of such functional block layouts can be developed, circuit modules that can be used in product after product. They feature fully debugged circuitry, a known history, and successful physical layout.

Common-impedance coupling in analog circuits having high sensitivity may result from sensors producing useful low-level signal outputs. Several solutions to the problem of common-impedance coupling exist, but only one or two may be viable in any given situation. For example,

- Increase conductor width and/or thickness
- Add high-frequency decoupling capacitors
- Use raised power-bus distribution schemes
- Use power supply and power and signal return planes

References

1. "Design Guidelines for Electronic Packaging Utilizing High-Speed Techniques," IPC-D-317, Proposal, August 1989, The Institute for Interconnecting and Packaging Electronic Circuits, Lincolnwood, Ill.
2. Joseph Smith, Steven F. Mango, Brian D. Morrison, Raytheon Co., "SPICE Modeling and Circuit Effects Extraction for High-Bandwidth PWB Design," *IPC Technical Review*, November 1986, pp. 13–20.
3. "Design Standard for Rigid Single- and Double-Sided Printed Boards," ANSI/IPC-D-319, January 1987, The Institute for Interconnecting and Packaging Electronic Circuits, Lincolnwood, Ill.
4. W. S. Fujitsubo, Hughes Aircraft Company, "Controlled Impedance Intercon-

nections: Theories, Problems, and Applications—Part 1," *Electri-Onics*, November 1986, pp. 56–58.

5. Charles A. Harper and William W. Staley, Technology Seminars Inc., "Interconnection Challenges with High Speed Digital Processing Systems," *Electronic Packaging & Production*, April 1985, pp. 58–62.

6. Ronald Pound, "Maintain the Speed of GaAs in Digital Systems," *Electronic Packaging & Production*, October 1986, pp. 50–52.

7. W. S. Fujitsubo, Hughes Aircraft Company, "Controlled Impedance Interconnections: Theories, Problems, and Applications—Part 2," *Electri-Onics*, December 1986, pp. 55–56.

8. Howard W. Markstein, "Packaging for High-Speed Logic," *Electronic Packaging & Production*, September 1987, pp. 48–50.

9. Mark Saubert and Dan Snyder, Elco Corp., "Packaging Engineers Face Conflicting Demands," *Electronic Packaging & Production*, June 1985, pp. 152–153.

10. Evan E. Davidson, IBM Corp., "Electrical Design of a High Speed Computer Packaging System," *IBM Journal of Research and Development*, vol. 26, no. 3, May 1982, pp. 349–361.

11. Stephen J. Ferry, Hughes Aircraft Co., "Microwave Boards from the Circuit Designer's Point of View", *Printed Circuit Design*, November 1984, pp. 3–7.

12. Tom Williamson, Intel Corp., "Minimizing Noise in Digital PCB Layouts," *Machine Design*, September 6, 1984, pp. 144–146.

13. Gil Condon, Condon & Associates, "Solve Two-Sided Printed Board EMI/Noise Problems the Easy Way—at the Drawing Board," *ITEM Update*, 1988.

14. Michael F. Violette, J. L. Norman Violette, JLN Violette & Associates, "EMI Control in the Design and Layout of Printed Circuit Boards," *EMC Technology*, March-April 1986, pp. 19–32.

15. Donald R. J. White, Don White Consultants, "EMI Control in the Design of Printed Circuit Boards," *EMC Technology*, January 1982, pp. 74–83.

16. John Kneubuhl, The Grass Valley Group, "Analog Circuit Design," *Printed Circuit Design*, November 1988, pp. 38–43.

Through-Hole Mounting

3.1 Introduction[1]

The most significant advantage of using the through-the-board mounting method is its compatibility with conventional mass soldering techniques, such as dip and wave soldering. Parts and components should be mounted on the side of the printed board opposite to the one that would be in contact with the solder if the board is machine-soldered. Except when mounted in cordwood modules or on non-repairable printed boards, parts and components should be so spaced and so located that any part can be removed from the printed board without removing another part. The following criteria and illustrations are examples of the information provided in ANSI/IPC-CM-770 for the design considerations associated with the through-hole mounting of electronic circuit components.

3.2 Component Mounting Considerations

The selection of a particular method for the through-hole mounting and connecting components in equipment will depend on the type of component package involved; on the equipment available for mounting and interconnecting; on the connection method used (soldered, welded, crimped, etc.); on the size, shape, and weight of the equipment package; on the degree of reliability and maintainability (ease of replacement) required; and, of course, on cost considerations.

3.2.1 Component type

Components are usually selected for electrical, thermal, or mechanical characteristics that are determined by the requirements of the contents of the package. The material composition, finish, and configuration of both the body and the terminations of a component must be considered in the choice of assembly methods. Table 3.1 contains a summary of the body and termination characteristics of the major classes of component to be assembled to printed boards.

All components must be qualified for the assembly processes to be used. The physical dimensions of the component must adequately mate with the physical handling devices of the placement equipment. The parts must not be degraded, physically or electrically, by soldering or other high-temperature processes used. Also, the parts must be able to tolerate exposure to the chemicals used in adhesive bonding, soldering, cleaning, or any other chemical processing.*

3.2.2 Component placement[3]

The complexity and variety of components available demands that proper component layout and placement procedures be followed to assure an economical design for production of printed circuit board assemblies. Therefore, when designing printed wiring assemblies, the following basic concepts should be considered:

- Specific contractual requirements.
- Select standard parts, materials, and processes.
- Design for use of standard tools.
- Safety factors for the equipment.
- Accessibility of parts.
- Minimum need for adjustment in operating the equipment.
- Provision for simple maintenance.
- Select parts resistant to damage from heat of soldering. Parts to be mounted must not only meet the storage and operating conditions specified for the equipment, but they must also maintain their integrity after having been subjected (during assembly) to heat soldering.

3.2.3 Lead configuration

The objectives of lead termination are to form the lead and electrically connect it to the conductors in such a manner that the required circuit continuity is provided through the life of the equipment regardless of

---

*For more information on through-hole mounting component types see Chap. 8 of McGraw-Hill's *Printed Circuits Handbook*.[2]

TABLE 3.1 Component Physical Characteristics

Component type	No. of leads	Body material	Organization	Termination Composition	Termination Finish
Axial-leaded	2	Plastic/composition	Linear	Metal lead	Solder
Radial-leaded	2	Plastic/composition	Linear	Metal lead	Solder
Chip component	2–4	Plastic/ceramic	Linear	Metalized end cap	Solder/gold/Ag-Pd
				Ribbon lead	solder/gold
Multileaded radial	3–12	Plastic/metal can	Perimeter	Metal lead	Solder
SOT	3–4	Post-molded plastic	Linear	Formed metal lead	Solder/silver
SOIC	4–28(40)	Post-molded plastic	Linear	Formed metal lead	Solder/silver
SIP	4–34	Plastic/metal/ceramic	Linear	Metal lead	Solder/gold
DIP	4–68	Post-molded/plastic/ceramic	Linear	Formed metal lead	Solder/gold
Ribbon-leaded	2–50	Plastic	Perimeter	Metal lead	Solder/gold
Leaded chip carrier	18–164	Plastic/ceramic	Perimeter	Formed metal lead	Solder/gold
Leadless chip carrier	18–164	Plastic/ceramic	Perimeter	Metalized pad	Solder/gold
Pin grid array	100+	Plastic/ceramic	Area	Metal lead	Solder
Pad grid array	100+	Plastic/ceramic	Area	Metalized pad	Solder/gold
Connector	100+	Plastic	Linear	Metal lead	Solder
Socket	100+	Plastic	All	Metal lead	Solder

SOURCE: ANSI/IPC-CM-770, Ref. 1.

95

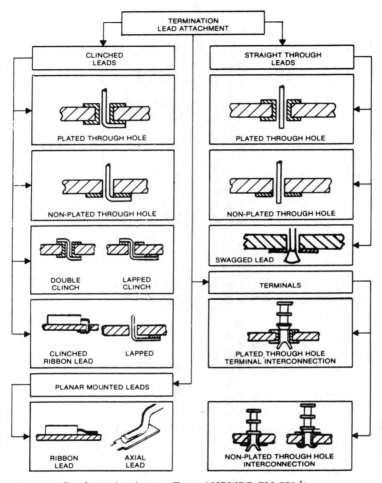

Figure 3.1 Lead terminations. (*From ANSI/IPC-CM-770.*[1])

the environments to which the assembly may be subjected. See Fig. 3.1 for the most frequently used methods.

Leads may be attached to unsupported printed conductor lands by clinching or straight-through (unclinched) lead attachment. The attachment should be completed by soldering. Component attachment to printed circuit boards should have the lead or terminal pass through the board and be soldered to the conductor pattern on the opposite side of the board. Lead attachment should normally be an option as to whether clinched or straight-through attachment is used, with the following restrictions:

- Where flat swaged eyelets (unfused) are used, a clinched lead attachment should be used.

- For straight-through attachment the diameter of unsupported holes should not exceed the diameter of the inserted lead by more than 0.5 mm (0.020 in).

- For straight-through lead attachment to supported holes the inside diameter of the supported hole should not exceed the diameter of the inserted lead by more than 0.7 mm (0.028 in).

- Leads should be terminated in such a manner that they do not exert a lifting force on the copper foil terminal area or conductor.

- Each functional lead should have an associated terminal area.

- There should be no more than one lead in any lead mounting hole.

3.2.3.1 Unclinched leads. The most direct method for mounting components to the printed circuit board is the straight-through method with unclinched leads. The use of straight-through unclinched leads requires the minimum of device handling, a straightening of the component leads, and cutting the leads to length before or after insertion.

Disadvantages. The disadvantages associated with this approach are

- The device is subject to movement both before and during the soldering operation. This makes it difficult to control the component height off the mounting surface. This movement can be a source of solder joint problems.

- It is difficult to maintain a suitable clearance between the body of the component and the printed circuit board surface for flux removal and, when applicable, conformal coating of the assembly. This is greatly minimized when multiple lead cans with integral standoffs are used, creating a component seating plane below the surface of the can from which the leads emerge.

- When the leads are rigid, precise drilling of the component mounting hole pattern is required due to small lead circle required and the inflexibility of the unformed leads.

- Supported holes are preferred in the printed circuit board for the component lead holes to enhance the mechanical strength of the solder joint. Otherwise the clearance between the component lead, the hole, and the circumscribing land must consider the lead-to-hole ratio and the hole-to-land difference that would allow sufficient remaining conductor to promote solderability.

■ The automatic insertion of the device leads in limited space can present problems.

The mounting of multilead component cans with plastic spacers has been used to overcome some of the disadvantages for the more conventional straight-through lead mounting techniques. Spacers with protrusions on one side should be mounted with the protrusions against the board.

Advantages. In addition to the considerations common to the straight-through mounting techniques, the mounting of component with spacers has the advantages of:

■ A suitable clearance between the component body and the printed circuit board can be maintained to facilitate soldering flux removal and conformal coating.

■ A bearing surface for the body is provided if the component leads are to be clinched.

■ The extension of unclinched leads beyond the printed circuit board surface can be more accurately controlled.

■ The height of the component body above the printed circuit board surface can be more accurately controlled; this is especially important when the printed circuit board assemblies are closely spaced.

■ The spacer helps to reduce the magnitude of mechanical stresses that are transmitted to the lead-body interface seal.

■ The lead mounting hole pattern need not be held as accurately as for unprepositioned component leads.

3.2.3.2 Clinched leads. Clinching of leads prior to soldering is commonplace, either as part of machine insertion or following hand insertion. The substrate land configuration and spacing to adjacent lands must be considered. Clinching in line with traces is good practice and trimming of leads before clinching is recommended where clinch direction may cause shorting to adjacent lands. It is generally not felt to be necessary to clinch all leads of a multileaded device unless required by the customer and equipment class.

The lead is passed through the board, clinched to make contact with the land or conductor, and then soldered. The lead or terminal should make contact with the conductor pattern before soldering. Leads should not extend beyond the edge of their lands; however, if overlap does occur, the lead should never violate electrical spacing requirements. The lead termination hole may be supported by eyelets or plated through holes or it may be unsupported. As its name implies, the component leads for this method are clinched to the printed circuit board land after they have passed through the lead hole.

Advantages. In addition to the considerations common to all straight-through mounting methods, this method has the following advantages:

- A reinforced mounting hole is not required; teardrop and offset lands can be used.
- This method does afford some resistance to movement during soldering.

Disadvantages. This method has some of the disadvantages mentioned for unclinched straight-through mounting, in addition

- Care must be taken when cutting the lead to length and forming the clinch to assure that minimum conductor clearances are provided when the clinched lead overhangs the land.
- The lead clinching operation, if not controlled properly, can unduly stress the component lead-to-can body seal.

3.2.4 Lead-Hole Relationships

The lead-to-hole clearance must be such as to provide for good soldering conditions. Generally 0.25 mm (0.010 in) to 0.5 mm (0.020 in) clearance in diameter is used. If the clearance is too small or too large, adequate wicking of solder does not result. A minimum protrusion through the substrate is often specified. The maximum is dependent on specific later process equipment used and end product design clearances. For rectangular leads the dimension across the diagonal should be considered as being the lead diameter.

3.2.4.1 Unsupported holes. In determining the difference between the diameter of an unsupported hole and that of the lead to be placed in the hole, the hole should be from 0.25 to 0.5 mm (0.010 to 0.020 in) larger than the lead diameter.

3.2.4.2 Supported holes. In determining the difference between the diameter of a supported hole and that of the lead to be placed in the hole, the hole should be from 0.25 to 0.7 mm (0.010 to 0.028 in) larger than the land diameter.

3.2.5 Component securing

Forming of leads for through-hole mounting components serves many purposes. These include the desire to retain the component in the substrate during subsequent handling prior to soldering or to provide a standoff. Such bends or loops are provided by tooling with some forming machines and/or can be introduced by hand-forming. Care must be exercised such that stresses are not introduced to the component or leads or to the solder joint during solidification.

The shock and vibration to which printed circuit board components are subjected during normal handling and environmental testing and use can damage the lead terminations and lead-to-component body seals. For this reason, many components should be mechanically secured to the mounting base. The more commonly used component securing methods are clips, clamps, and brackets; wire and elastic straps; adhesives; and integral mounting provisions.

Most circuit malfunctions in a severe vibration environment are caused by cracked solder joints, cracked seals, or broken electrical lead wires. These failures are usually due to dynamic stresses that develop because of relative motion between the electronic components and the board. This relative motion is generally most severe during resonant conditions.

Since shock and vibration vary widely with the specific application, it is not possible to provide solutions to all component mounting problems encountered in every combination of environmental stress likely to be encountered. The purpose is to suggest some general guidelines which if observed will provide reasonable assurance that the components and assemblies will survive shock and vibration within their intended use. The extent to which the user wishes to implement these guidelines may ultimately be validated by actual tests of the assembled printed circuit board in its intended shock and vibration environment.

The ultimate ability of components to survive in shock and vibration environments will depend upon the degree of consideration given to the following factors:

- The worst-case levels of shock and vibration environment for the entire structure in which the printed circuit board assembly resides and the ultimate level of this environment that is actually transmitted to the components mounted on the board.

- Particular attention should be given to equipment which will be subjected to random vibration.

- The method of mounting the board in the equipment to reduce the effects of this environment, specifically the number of board mounting supports and their interval and complexity.

- The attention given to the mechanical design of the board; specifically its size, shape, type of material, material thickness, and degree of resistance to bowing and flexing that the design provides.

- The shape, mass, and location of the components mounted on the board.

- The component lead wire strain relief design as provided by its package, lead spacing, lead bending, or a combination of these plus the addition of restraining devices.

- The attention paid to workmanship during board assembly so as to ensure that component leads are properly bent, not nicked, and that components are installed in a manner which minimizes component movement.

3.2.6 Automated assembly

With few exceptions, automated techniques and equipment are now available for most component types and assemblies. This equipment ranges from dedicated devices for specific component types (radial leaded, axial leaded, chip carriers, discrete chips, sockets, etc.) to flexible "robotic" equipment programmed to handle a range of parts. Use of these techniques requires guidelines for design such as those described in Chap. 5 in component layout, orientation, spacing, and component standardization. These factors also aid manual assembly, but deviations are more easily accommodated with manual techniques.

Clearance for tooling "footprints" is required. Hole tolerances for through-hole components and land size tolerances for surface-mounted components are important and are generally smaller than those necessary with manual techniques. Sequence of component assembly is of more concern with automated techniques, as is clearance from substrate edges. Layout of traces to prevent bridging as a result of clinching must be considered.

Component tolerances, packaging and lead materials are important factors in assuring repeatable results with machine insertion. Change notice volume can result in high indirect support requirements for automated assembly. In addition to insertion rates in the thousands per hour, accuracy and repeatability of machine-controlled assembly surpasses that of manual assembly. Resultant process control and increased quality is more easily achieved for either high-mix low-volume or high-volume assembly.

One of the major advantages of using printed circuit boards is its suitability for use with numerically controlled automatic component insertion equipment. In order to obtain the manufacturing cost savings of using this type of equipment, several printed circuit design parameters must be taken into account which are not present when the printed circuit board is manually assembled.

3.3   Axial-Leaded Discrete Components

Axial-leaded components with two leads are perhaps the most common electrical components used in printed circuit assemblies. The component body is usually cylindrical in shape with two leads exiting from the opposite ends of the component along its longitudinal axis. The lead is usually round in cross section. Component identification as well as polarity, when necessary, are generally marked on the body of the component. Many resistors, capacitors and diodes are supplied in this configuration.

3.3.1   Component preparation

Component preparation is the processing step which generally includes forming and cutting of component leads to facilitate subsequent component assembly and/or minimize component damage due to stress. The lead should extend approximately one forming allowance straight out from the body of the component. This forming allowance is usually expressed as "2 lead diameters" or a minimum of 1.5 mm (0.06 in) prior to the start of the bend except that when space is limited by high-density packaging the minimum may be 0.75 mm (0.03 in). The end of the body in this application is defined to include any coating meniscus, solder seal, solder or weld bead, or any other extension.

The minimum component center-to-center board lead spacing can be represented Fig. 3.2)

$$L = B_{max} + 3D + 2FA$$

where L = center-to-center lead spacing
B = body length

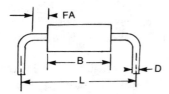

Figure 3.2 Axial-leaded component spacing criteria. (*From ANSI/IPC-CM-770.*[1])

D = nominal lead diameter[*]

FA = forming allowance (lead should not be disturbed within this distance from the body)

The value of L is usually adjusted upward to coincide with the grid used. The total length of both leads should not exceed 25 mm (1 in) in length unless this component is mechanically supported to the mounting base.

Recommended center-to-center lead spacing values for typical axial-leaded discrete components are given in Tables 3.2 (resistors), 3.3 (diodes) and 3.4 (capacitors).[4] The values in these tables may have to be increased when automatic insertion equipment is used. Where practical, components with similar physical dimensions should have the same lead spacing and be on the design grid.

3.3.2 Land patterns

Land patterns must consider minimum and maximum lead spacing requirements. Standard land spacing patterns should be established for the purpose of uniformity of assemblies and the practical use of assembly tools and equipment. Spacing(s) should be located to accommodate automatic assembly and "bed-of-nails" type of testing equipment [usually 2.5 mm (0.100 in) increments—i.e., 7.5 mm (0.300 in), 10 mm (0.400 in), 12.5 mm (0.500 inch), etc.].

Land patterns for unsupported holes should have more solderable area than that for supported holes for a stronger joint after soldering. The optimum dimension is dependent on the device and its mounting characteristics. The most common geometry is the round land with a centered hole. Square lands with centered holes are sometimes used to indicate polarity for polarized components.

3.3.3 Component placement

Through-hole components should be mounted on one side of the board only. Components should not be mounted across or on top of vias, exposed conductive patterns, and other components. Component bodies should not be closer than 1.5 mm (0.06 in) from the board edge. (See Fig. 3.3.)

3.3.3.1 Component orientation. Horizontally mounted components should be oriented in either of the two board axes preferably with component identification in the same direction. (See Fig. 3.4.) The body of the component should be approximately centered between the lead

---

[*]For lead diameters up to 0.7 mm (0.027 in). $4D$ for lead diameters between 0.7 and 1.2 mm (0.27 and 0.048 in) and $5D$ for lead diameters over 1.2 mm (0.048 in).

TABLE 3.2 Resistor Component Mounting Parameters and Recommended Lead Spacing, Inches

Style	Specification (ref.)	maximum body length B (ref.)	Nominal lead diameter D (ref.)	Lead spacing L		
				Class 1 0.025-in grid	Class 2 0.050-in grid	Class 3 0.100-in grid
RBR52	MIL-R-39005/1	1.020	0.032	1.200	1.200	1.200
RBR53	MIL-R-39005/2	0.770	0.032	0.950	0.950	1.000
RBR54	MIL-R-39005/3	0.770	0.032	0.950	0.950	1.000
RBR55	MIL-R-39005/4	0.520	0.032	0.700	0.700	0.700
RBR56	MIL-R-39005/5	0.364	0.032	0.525	0.550	0.600
RBR57	MIL-R-39005/7	1.020	0.032	1.200	1.200	1.200
RBR74	MIL-R-39005/8	0.520	0.025	0.675	0.700	0.700
RBR75	MIL-R-39005/9	0.315	0.025	0.475	0.500	0.500
RBR76	MIL-R-39005/10	0.832	0.032	1.000	1.000	1.000
RCR05	MIL-R-39008/4	0.160	0.016	0.275	0.300	0.300
RCR07	MIL-R-39008/1	0.281	0.025	0.425	0.450	0.500
RCR20	MIL-R-39008/2	0.116	0.032	0.600	0.600	0.600
RCR32	MIL-R-39008/3	0.593	0.040	0.800	0.800	0.800
RC42	MIL-R-11/7	0.728	0.045	0.950	0.950	1.000
RLR05	MIL-R-39017/5	0.170	0.016	0.300	0.300	0.300
RLR07	MIL-R-39017/1	0.281	0.025	0.425	0.450	0.500
RLR20	MIL-R-39017/2	0.416	0.032	0.600	0.600	0.600
RLR32	MIL-R-39017/3	0.593	0.040	0.800	0.800	0.800
RLR42	MIL-R-39017/4	0.728	0.045	0.950	0.950	1.000
RL42	MIL-R-22684/4	0.728	0.045	0.950	0.950	1.000
RNR50	MIL-R-55182/7	0.170	0.016	0.300	0.300	0.300
RNR55	MIL-R-55182/1	0.281	0.025	0.425	0.450	0.500
RNR65	MIL-R-55182/5	0.656	0.025	0.800	0.800	0.800
RNR70	MIL-R-55182/6	0.875	0.032	1.050	1.050	1.100
RN55	MIL-R-10509/7	0.281	0.025	0.425	0.450	0.500

RN60	MIL-R-10509/1	0.437	0.025	0.600	0.600	0.600
RN65	MIL-R-10509/2	0.656	0.025	0.800	0.800	0.800
RN70	MIL-R-10509/3	0.875	0.032	1.050	1.050	1.100
RN75	MIL-R-10509/5	1.124	0.032	1.300	1.300	1.300
RWR71	MIL-R-39007/5	0.874	0.032	1.050	1.050	1.100
RWR74	MIL-R-39007/6	0.937	0.040	1.125	1.150	1.200
RWR78	MIL-R-39007/7	1.842	0.040	2.050	2.050	2.100
RWR80	MIL-R-39007/8	0.437	0.020	0.575	0.600	0.600
RWR81	MIL-R-39007/9	0.281	0.020	0.425	0.450	0.500
RWR82	MIL-R-39007/12	0.328	0.020	0.500	0.500	0.500
RWR84	MIL-R-39007/10	0.937	0.040	1.125	1.150	1.200
RWR89	MIL-R-39007/11	0.622	0.032	0.800	0.800	0.800
RW67	MIL-R-26/4	1.094	0.036	1.275	1.300	1.300
RW68	MIL-R-26/4	1.938	0.036	2.125	2.150	2.200
RW69	MIL-R-26/4	0.563	0.036	0.750	0.750	0.800
RW70	MIL-R-26/5	0.437	0.020	0.575	0.600	0.600
RW79	MIL-R-26/5	0.622	0.032	0.800	0.800	0.800

Note: Wherever possible use the largest class number allowed by the application; 1 in = 2.54 cm.
SOURCE: From IPC-D-330, Ref. 3.

TABLE 3.3 Diode Component Mounting Parameters and Recommended Lead Spacing, Inches

Style	Specification (ref.)	Maximum body length B (ref.)	Nominal lead diameter (ref.)	Lead Spacing L		
				Class 1 0.025-in grid	Class 2 0.050-in grid	Class 3 0.100-in grid
1N3016 thru 1N3051	MIL-S-19500/115	0.560	0.032	0.725	0.750	0.800
1N3154 1N3155 1N3157	MIL-S-19500/158	0.300	0.020	0.425	0.450	0.500
1N3600	MIL-S-19500/231	0.300	0.020	0.425	0.450	0.500
1N3821 thru 1N3828	MIL-S-19500/115	0.560	0.032	0.725	0.750	0.800
1N4148	MIL-S-19500/116	0.180	0.020	0.325	0.350	0.400
1N4245 thru 1N4249	MIL-S-19500/286	0.300	0.032	0.475	0.500	0.500
1N4370 thru 1N4372	MIL-S-19500/127	0.300	0.020	0.425	0.450	0.500
1N4946 thru 1N4948	MIL-S-19500/359	0.560	0.032	0.725	0.750	0.800
1N4954 thru 1N4996	MIL-S-19500/356	0.350	0.040	0.550	0.550	0.600
1N645 1N647	MIL-S-19500/240	0.300	0.020	0.425	0.450	0.500
1N746 thru 1N759	MIL-S-19500/127	0.300	0.020	0.425	0.450	0.500
1N821 1N823 1N825 1N829	MIL-S-19500/159	0.300	0.020	0.425	0.450	0.500
1N914	MIL-S-19500/116	0.180	0.020	0.325	0.350	0.400
1N935 1N938 1N939	MIL-S-19500/156	0.300	0.020	0.425	0.450	0.500
1N941 1N944 1N945	MIL-S-19500/157	0.300	0.020	0.425	0.450	0.500
1N962 thru 1N992	MIL-S-19500/117	0.300	0.020	0.425	0.450	0.500

Note: Wherever possible use the largest class number allowed by the application; 1 in = 2.54 cm.
SOURCE: From IPC-D-330, Ref. 4.

TABLE 3.4   Capacitor Component Mounting Parameters and Recommended Lead
Spacing, inches

Specification (ref.)	Maximum body length B (ref.)	Nominal lead diameter D (ref.)	Lead spacing L		
			Class 1 0.025- in grid	Class 2 0.050- in grid	Class 3 0.100- in grid
MIL-C-39006	0.828	0.025	.975	1.000	1.000
(Style CLR)	1.016	0.025	1.175	1.200	1.200
	1.141	0.025	1.300	1.300	1.300
	1.469	0.025	1.625	1.650	1.700
	1.625	0.025	1.775	1.800	1.800
	1.938	0.025	2.100	2.100	2.100
	2.625	0.025	2.775	2.800	2.800
	3.250	0.025	3.400	3.400	3.400
MIL-C-39003/1	0.422	0.020	0.550	0.550	0.600
(Style CSR13)	0.610	0.020	0.750	0.750	0.800
	0.822	0.025	0.975	1.000	1.000
	0.922	0.025	1.075	1.100	1.100
MIL-C-39014/5	0.190	0.025	0.350	0.350	0.400
(Style CKR11)					
MIL-C-39014/5	0.280	0.025	0.425	0.450	0.500
(Style CKR12)					
MIL-C-39014/5	0.420	0.025	0.575	0.600	0.600
(Style CKR14)					
MIL-C-39014/5	0.540	0.025	0.700	0.700	0.700
(Style CKR15)					
MIL-C-39014/5	0.740	0.025	0.900	0.900	0.900
(Style CKR16)					
MIL-C-39022/1	1.094	0.020	1.225	1.250	1.300
(Style CHR09)	1.156	0.020	1.300	1.300	1.300
	1.218	0.025	1.375	1.400	1.400
	1.281	0.025	1.425	1.450	1.500
	1.468	0.025	1.625	1.650	1.700
	1.531	0.025	1.675	1.700	1.700
	1.781	0.032	1.950	1.950	2.000
	1.968	0.032	2.150	2.150	2.200
	2.156	0.032	2.325	2.350	2.400

NOTE: Wherever possible use the largest class number allowed by the application; 1
in = 2.54 cm.
SOURCE: From IPC-D-330, Ref. 5.

spacing. In general, components with cylindrical bodies should be in
direct contact to the mounting base. Density of component mount
should consider clearance needed for component assembly, electrical
clearance, and ease of rework.

3.3.3.2 Vertical mounting.
Components mounted perpendicular to
printed circuit boards should be installed with a minimum of 0.4 mm
(0.016 in) clearance between the end of the component body, which in-

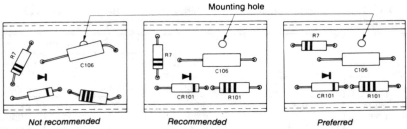

Mounting hole

Not recommended   *Recommended*   *Preferred*

Figure 3.3 Alignment boundaries. (*From ANSI/IPC-CM-770.*[1])

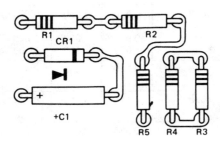

Figure 3.4 Horizontally-mounted axial-leaded components. (*From ANSI/IPC-CM-770.*[1])

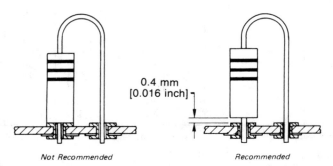

0.4 mm
[0.016 inch]

Not Recommended   *Recommended*

Component mounted flush to printed board.   Component mounted with adequate clearance between end of body and board; spacer may be used.

NOTES:

1. The purpose of raising the component body off the board surface is to facilitate solder flow, escape of gases in soldering and cleaning. Normally, this is accomplished only by raising the component off the board. There are, however, certain hole support devices that can support the component and still accomplish solder flow, gas liberation, and cleaning.

2. If the component is mounted tight to the board (no clearance) component movement will unduly stress the component lead.

3. Conformal coating application and coverage may necessitate an increase in the 0.015" minimum dimension.

4. Care should be taken when using spacers with high coefficients of thermal expansion.

Figure 3.5 Vertically-mounted axial-leaded components. (*From ANSI/IPC-CM-770.*[1])

cludes any packaging meniscus and the surface of the board, to prevent potential heat damage and entrapment problems. (See Fig. 3.5.)

3.4 Radial-Leaded Discrete Components

Radial-leaded components come in a variety of shapes: cylindrical, square, rectangular, wafer, and kidney. Leads exit from a common side of the component. The lead is either ribbon-shaped or cylindrical. Selected devices can be automatically inserted during assembly and also can be modified with coined leads for surface mounting. Typical types of radial-lead components are electrolytic, plastic, dipped, molded, and encapsulated capacitors and transistors.

3.4.1 Component preparation

Under certain circumstances, it may be advisable to provide stress relief for the leads of radial-lead components by forming the leads outward from the normal component lead pattern to an enlarged pattern to provide a degree of stress relief. Another method of providing stress relief for radial-lead components is the use of a flexible low-modulus spacer supported by rounded protrusions.

3.4.2 Land patterns

Information on land patterns follows that cited in Sec. 3.3 for axial-leaded components.

3.4.3 Component placement

3.4.3.1 Vertical mounting. Components whose leads are on one side should be mounted so their axes are as parallel and perpendicular to the mounting base as practicable. Components should have their vertical axis within 15° of the perpendicular plane of the mounting base unless restricted by design. Any bend on the lead should not fracture the case.

The coating on leads should not be removed beyond where the lead enters the component body. Components should be placed so that the coating does not enter the mounting hole. Components installed on boards with lands on the component side should maintain a minimum clearance of 0.4 mm (0.016 in) between the surface of the circuitry and where the coating ends on the lead.

3.4.3.2 Horizontal mounting. When components are mounted horizontally, mounting clips or adhesives should be used. A typical method of mounting a component of this type is shown in Fig. 3.6.

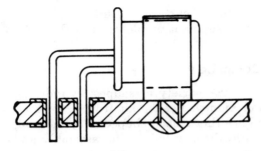

Figure 3.6 Horizontally-mounted radial-leaded component. (*From ANSI/IPC-CM-770.*[1])

3.5   Multiple-Radial-Lead Components

The packaging technology is well established for transistors in metal "TO" cans. This configuration was used for early multiple-lead components and is still popular today. Multiple-lead component cans are also available in many sizes and shapes. The following general considerations should be taken into account when designing printed circuit board assemblies with multiple-lead components:

- Land size, lead forming, and lead clinching
- The physical dimensions of the multiple-lead component
- Automatic, semiautomatic, and manual component insertion tolerances and restraints
- Component dimensions and tolerances
- Mechanical securing such as clips, clamps, brackets, and sockets

This type of component consists of a hermetically sealed can with up to 12 round leads exiting from the bottom of the device (usually in a circular pattern). Dimensions of standard and registered TO devices are included in JEDEC 95.[5] Available tooling, hermetic sealing, and a rugged construction made the can with 10 or 12 leads a natural first IC package. It requires special punching dies, drilling templates or off-grid numerically controlled (NC) drill programming for the 5.84-mm (0.230-in) diameter pin circle.

3.5.1   Component preparation

Multiple-lead cans are used in many printed circuit board applications. However, the wide variety of multiple-lead can sizes and number of leads make it impossible to standardize on a mounting method for such devices. Leads exiting from multilead radial-type components may be formed to standard grid spacings and out away from under the body of the component. This technique is used to provide inspection of solder joints, stress-relieving component leads, enhanced cleaning, etc.

3.5.2   Land patterns

Land patterns for multilead radial packages are a function of the dimensions and number of leads as they exit the body of the component to be mounted. Pattern configuration will also vary depending upon the lead-forming requirements, as with spreader mounting or reform into alternate patterns.

In Fig. 3.7 the leads of a TO-100 are formed to a standard 6.4 × 9.5 cm (0.250 × 0.375 in) pattern, leaving space for two via or plated holes available for internal plane connections or for conductor routing. Lead forming provides a standoff, and protective coating is specified for the carrier surface to insulate the printed wiring.

3.6   Dual–Inline Packages (DIPs)[6]

The DIP multiple lead component has its leads pointing downward, ready for insertion into holes in a printed circuit board. DIP layouts usually follow a square grid pattern and reasonable densities are possible, such as that shown in Fig. 3.8. Registered dual–inline outlines are described in JEDEC-95[5] and Chap. 5.

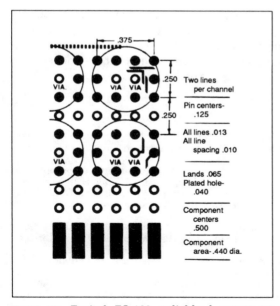

Figure 3.7 Typical TO-100 radial-lead component layout. (*From ANSI/IPC-CM-770.*[1])

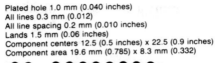

Plated hole 1.0 mm (0.040 inches)
All lines 0.3 mm (0.012)
All line spacing 0.2 mm (0.010 inches)
Lands 1.5 mm (0.06 inches)
Component centers 12.5 (0.5 inches) x 22.5 (0.9 inches)
Component area 19.6 mm (0.785) x 8.3 mm (0.332)

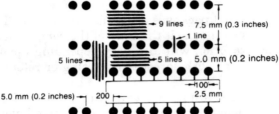

Figure 3.8 Typical dual–in–line package (DIP) layout. (*From ANSI/IPC-CM-770.*[1])

A DIP is made of metal, ceramic, glass, plastic, or combinations of these materials. Leads, body, and glass sealing are designed to make this a rugged package. The shape and forming of the leads provides both a standoff and an interference fit in the component mounting holes. This feature eliminates lead bending to ensure retention, and eases removal of the unit. Leads are 2.54 mm (0.100 in) apart, center to center, which restricts artwork. If the cost of small hole and land sizes can be justified, a large number of printed wires per channel can be obtained with the DIP.

The 14-lead TO-116 DIP can be inserted easily, either automatically or manually. Lead flagging and clinching are not required, owing to the interference fit. For hand assembly, hand tools or fixtures are required, and automatic insertion equipment is available. Sockets are available for mounting and testing purposes.

References

1. "Printed Board Component Mounting," ANSI/IPC-CM-770, Revision C, February 1987, Institute for Interconnecting and Packaging Electronic Circuits, Lincolnwood, Ill.
2. Gerald L. Ginsberg, Component Data Associates, Inc., "Circuit Components and Hardware," Chapter 8, *Printed Circuits Handbook, 3d* ed., McGraw-Hill, New York.
3. "Component Placement," IPC-D-330, No. 7.2, "Printed Wiring Design Guide," May 1972, Institute for Interconnecting and Packaging Electronic Circuits, Lincolnwood, Ill.
4. "Axial-Lead Components," IPC-D-330, No. 7.1.1, "Printed Wiring Design Guide," May 1972, Institute for Interconnecting and Packaging Electronic Circuits, Lincolnwood, Ill.
5. "JEDEC Registered and Standard Outlines for Solid State Products," Publication 95, Electronic Industries Association, Washington, D.C.
6. "Multiple Lead Components," IPC-D-330, No. 7.1.3, "Printed Wiring Design Guide," May 1972, Institute for Interconnecting and Packaging Electronic Circuits, Lincolnwood, Ill.

Surface Mounting

4.1 Introduction[1,2]

Surface-mount technology (SMT) presents distinct advantages to the manufacturer who needs to reduce product size or increase product functionality. However, without adhering to proper design rules, these advantages are quickly lost in assembly, rework, quality assurance (QA), and testing. Today's professional designer must work closely with engineering, manufacturing, purchasing, testing, and QA. Together this team, with a good foundation in SMT design and manufacturing techniques, can eliminate most assembly problems, including tombstoning, component shifting, or cracking during the reflow process.

When designing for surface-mount component assemblies, several factors must be considered, including: the printed circuit board and how it will be constructed and spaced; the manufacturing equipment and its positioning and attaching requirements; soldering and cleaning processes and their impact and success on both components and printed circuit board; and finally, the testing fixtures and methods. Since standards sometimes vary from one component manufacturer to another, it is imperative that the designer work closely with the component engineer and purchasing to ensure that the correct component is designed into the product. These are the same concerns associated with designing for standard assembly, but in each area, a number of significant differences exist between conventional through-hole and surface-mount assembly. Also, the combination of the two component-mounting methods present additional requirements in design.

4.2 Fabrication Panel Considerations[3]

In order to fully utilize the automation technology associated with surface-mount components, a designer should consider how a board or packaging and interconnecting (P&I) structure will be manufactured, assembled, and tested. Each of these processes, because of the particular equipment used, requires fixturing, which will affect or dictate certain facets of the board layout. Tooling holes, panel size, component orientation, and clearance areas (both component and conductor) on the primary and secondary sides of the board are all equipment- and process-dependent.

To produce a cost-effective layout through optimum base material utilization, a designer should consult with the manufacturer to determine typical panel size. The board should be designed to utilize the manufacturer's suggested usable area. Smaller boards can be ganged on this same panel size to simplify fixturing and reduce further handling. Most manufacturers will suggest various methods of retaining assemblies in panels. A method should be chosen that takes the assembly and test processes into consideration. (See IPC-D-322 for panel to board relationships.[4])

Small boards can effectively be arranged on a single working panel through all steps of the manufacturing process if the designer works closely with manufacturing. The specific limitations and requirements of solder screening, component placement, soldering, and test equipment used should be established. Equipment manufacturers may also provide this information.

4.2.1 Copper distribution

There are various items which designers have control over which induce problems after fabrication. For example the designer should try to make sure that heavy copper planes are located equidistant from the neutral axis. If not, bowing due to bimetallic action will occur after the board or P&I structure is laminated. Test pattern should be generated external to artwork so that manufacturing variables can be verified without destroying a board. This is needed primarily for multilayer boards.

All boards have internal fibers or reinforcement. Because of weaving, these are stressed differently. Within a multilayer design, it is possible to alternate the layers. This will aid in reducing bow. Plates used to restrain the material during lamination with rigid pins induce stresses which encourage buckling; this in turn results in waviness in boards. Boards with very high resin to fiber ratios magnify internal copper trace concentrations, which may show up as flatness deviations. Areas where boards are to be routed should retain copper foil. This will prevent ragged edges due to fiber unraveling.

Some designs require thermal lands beneath chip carrier elements. They are often needed to dissipate the heat from critical components.

4.2.2 Panel format

Components can be mounted on individual boards or on boards that are still organized in panel form. Boards or panels that are to be handled by automatic board-handling equipment or are to pass through automated parts placement, soldering, cleaning, etc., steps must have areas along the sides kept free of parts or active circuitry. Special tooling and fixturing holes are generally located within the edge clearance areas. The clearance areas are needed to avoid interference with board-handling fixtures, guidance rails, and alignment tools.

Typically a strip of width 3.8 to 10 mm (0.150 to 0.400 in) must be allowed along the sides for the clearance. The required clearance width is dependent upon the design of the board-handling and fixturing equipment. These dimensions should be obtained from the equipment manufacturer before board or panel design.

For accurate fixturing, a minimum of two (and preferably four) nonplated holes are located in the corners of the board to provide accurate mechanical registration on parts placement equipment. The diameters of the holes are typically between 2.5 and 3.8 mm (0.100 and 0.140 in). Specific sizes should be obtained from the equipment manufacturer. Also, optical fiducial marks may be located near the fixturing holes if optical alignment is used to improve registration.

Board-handling holes of 3.17 to 6.35 mm (0.125 to 0.250 in) in diameter may also be located in the clearance areas. These holes are used by automated board-handling equipment to move boards (or panels) from station to station in automated assembly lines.

Printed board designs implementing SMT components should include optic targets to orient and register the component lands to the center of the device and/or the artwork image to the P&I structure. This may be accomplished by using three fiducials placed on a known grid. The same grid can be used as the origin (center point of body or designed pin no. 1) for all surface-mount components. This will allow the automatic placement machines to use optic positioning equipment and methods. These same fiducials will also allow the use of vision inspection equipment and methods in P&I fabrication and assembly. Such fiducials serve many of the same functions as tooling holes in through-hole printed board technology. Both are normally required on surface-mount printed circuit board designs.

The fiducial size and shape is normally dependent on the various vision systems. Three of the fiducials should be placed on grid in the corners of each printed board to form a three-point datum system. The solder mask opening should be a larger diameter than the fiducial diameter.

4.3   Component Considerations[1,2]

At an early point in the design cycle, a decision will have to be made as to which component types will be used in the assembly. As the components will significantly influence the final printed circuit board design, this bill of materials should be reviewed carefully. All second sources and other possible alternatives should be listed and reviewed at this stage because redesign at later stages will become more expensive.

Component selection factors will include package configuration, component lead styles, component profile and footprint patterns, and also, component availability must be confirmed and the documentation supplied to the designer. (For more information on surface-mount component selection see Sec. 4.7 and Chapter 3 of McGraw-Hill's *Printed Circuit Handbook*)[6].

4.3.1   Substrate Population[7]

Population density of surface-mount components over the total area of the substrate must also be carefully considered, as placement machine limitations can create a "lane" or "zone" that restricts the total number of components which can be placed within that area on the substrate. For example, on a hardware-programmable simultaneous placement machine, each pick-and-place unit within the placement module can only place a component on the substrate in a restricted lane (owing to adjacent pick-and-place units), typically 10 to 12 mm (0.4 inch to 0.48 inch) wide. Thus, the placement of 10 components in a lane will require a machine with 10 placement modules (or 10 passes beneath a single placement module), an inefficient process considering that there are no more than three surface-mount components in any other lane.

A special consideration for mixed printed circuit assembly technology designs in the location of leaded through-hole mount components with respect to the surface-mount components and the minimum distance between a protruding clinched lead and a conductor or surface-mount land pattern. Minimum distances between the clinched lead ends and the surface-mount components or substrate conductors should be 1 mm (0.04 in) and 0.5 mm (0.02 in), respectively.

4.3.2 Component spacing[7,8]

Of all the issues in design for manufacturability, component spacing is the most important one. It controls cost effectiveness of placement, soldering, testing, inspection, and repair. A minimum component spacing is required to satisfy these manufacturing requirements. There is no limit on maximum interpackage spacing, the more the better. Some designs require that surface-mount components be positioned as tightly as possible. This is not always a good practice.

The minimum component pitch is governed by the maximum width of the component and the minimum distance between adjacent components. When defining the maximum component width, the rotational accuracy of the placement machine must also be considered. The minimum permissible distance between adjacent surface-mount components is a fig ure based upon the gap required to avoid solder bridging during the wave soldering process. This distance plus the maximum component width are combined to derive the basic expression for calculating the minimum pitch.

As a guide, the recommended minimum pitches for various combinations of two sizes of surface-mount components, the R/C1206 and C0805 (R or C designating resistor or capacitor respectively, the number referring to the component size), are given in Table 4.1. The values are not based on worst-case conditions, but on a statistical analysis of all boundary conditions. There is a certain flexibility in the given data. For example, it is possible to position R/C1206 surface-mount components on a 2.5-mm (0.1-in) pitch, but the probability of component placements occurring with G_{min} smaller than 0.5 mm (0.020 in) will increase, hence the likelihood of solder bridging also increases. Each application must be assessed on individual merit, with regard to acceptable levels of rework and so on.

4.3.3 Component Orientation[3,8]

Another consideration for manufacturing is alignment of components on printed circuit boards. Similar types of components should be aligned in the same orientation for ease of component placement, inspection, and soldering. In wave soldering, proper alignment is necessary to prevent solder skips or bridging. On any printed circuit board assembly where the secondary side is to be wave-soldered, the preferred orientation of devices on that side should be such as to optimize the resulting solder joint quality as the assembly exits the solder wave. Thus

- All passive components should be parallel to each other.

TABLE 4.1   Typical Discrete Chip Component Spacing, mm (in)

Combination	Component A	Component B R/C1206	C0805
F_{min} (A above B)	R/C1206 C0805	3.0 (.12″) 2.8 (.112″)	2.8 (.112″) 2.6 (.104″)
F_{min} (A beside B)	R/C1206 C0805	5.8 (.232″) 5.3 (.212″)	5.3 (.212″) 4.8 (.192″)
F_{min} (A beside B vertical)	R/C1206 C0805	4.1 (.164″) 3.6 (.144″)	3.7 (.148″) 3.0 (.12″)

NOTES: These figures are statistically derived under certain assumed boundary conditions, as follows: Positioning error (Δp) ± 0.3 mm (± 0.012 in); pattern accuracy (Δq) ± 0.3 mm (± 0.012 in); rotational accuracy (ϕ) ± 3°; component metalization/solder land overlap (M_{min}) 0.1 mm (0.004 in) (note this figure is only valid for wave soldering); the figure for the minimum permissible gap between adjacent components in (G_{min}) is taken to be 0.5 mm (0.020 in).
SOURCE: North American Philips Corp.

- All small-outline integrated circuits (SOICs) should be parallel to each other.

- The longer axes of SOICs and of passive components should be perpendicular to each other.

- The long axes of passive components should be perpendicular to the direction of travel of the board along the conveyor of the wave soldering machine.

The orientation of surface-mounted components on printed board assemblies to be soldered by other production methods is less critical. In addition, orienting similar components in the same direction is very desirable for pick-and-place equipment because not all machines have head rotational capability. Uniform orientation is also very desirable for the inspecting component for misplacement.

4.4   Basic Board Features[3]

In general, the basic printed circuit board features for surface-mount design are smaller than those for through-hole technology. Additionally, more vias are often used to connect layers. These vias also are smaller than the holes used to hold component leads. Thus, for surface-mount technology the use of smaller conductors, spaces and vias complicate the design and printed circuit board fabrication process.

4.4.1   Conductor routing

Conductors connecting to a land area can act as solder thieves by drawing solder away from the land and down the conductor. To minimize this, several options are available, for example,

- Narrow the conductor as it enters the land area. Maximum conductor width should be 0.38 mm (0.015 in).

- Route conductors into the lands in a "necked-down" configuration. This prevents component swim.

The most recommended option is

- Use solder mask over lines that are bare copper or copper that has had the plating selectively removed. The solder mask and bare copper provide an effective barrier to solder migration. This may provide sufficient protection even if the other options are ignored.

4.4.2   Vias

Another consideration for manufacturing is selecting via sizes and locations for interconnection. The size of the via should be selected based upon the board thickness versus hole diameter or aspect ratio. Its location may be specified as a part of the component land pattern if the via is to be used for electrical testing or may be allowed to be located randomly if auxiliary test lands are provided in fixed locations.

4.4.2.1   Vias within component mounting lands.   Vias are normally plated-through holes in 0.63 to 1.0 mm (0.025 to 0.040 in) diameter lands. Unless properly treated, they must be located away from the component lands to prevent solder migration off the component land during reflow soldering. This will cause insufficient solder fillets on components. Solder drain can be prevented by providing a narrow

bridge between the land area and the via or by using the solder mask over bare copper circuitry method or its equivalent.

Specifying tented or filled vias will also reduce solder migration on assemblies with solder reflow circuitry and eliminate its possibility with nonmelting metal circuitry (i.e., solder mask over bare copper). Filled or tented vias also take care of potential flux entrapment problems under components and are highly desirable for attaining good vacuum seal during in-circuit bed-of-nails testing.

For wave soldering applications, vias within lands of passive components glued to the underside are acceptable or even preferred because they may provide escape holes for outgassing. This may prove to be especially helpful in case the flux used in wave soldering did not sufficiently dry during preheat.

4.4.2.2 Vias under components. Vias may be placed under surface-mounted components if they are to be reflow-soldered. However, if the assembly is to be wave-soldered, vias underneath components should be avoided or tented with solder mask because, during wave soldering of these assemblies, flux may potentially become trapped under these packages. For effective cleaning, vias may be located underneath surface mount packages in full surface-mount assemblies that will not be wave-soldered.

4.4.2.3 Vias as test points. Vias are used to connect surface-mounted component lands to conductor layers. They may also be used as test targets for bed-of-nails type probes and/or rework ports. Vias may be tented if they are not required for node testing or rework. When a via is used as a test point, it is required that the location of a test land be found to match the standard grid of the test fixture.

4.4.3 Conductors and vias

Figure 4.1 shows the typical space used to separate conductors, lands, and vias and the tolerances held in the printed circuit board fabrication. These spacings will vary depending on the method used for artwork generation [manual versus computer-aided design (CAD) generation]. Figures 4.2 and 4.3 show typical relationships between conductors and lands for two- and three-conductors-per-channel routing, respectively.

4.5 General Land Pattern Considerations[1,3]

Surface land patterns define the sites where the components are to be soldered to the printed circuit board or other packaging and intercon-

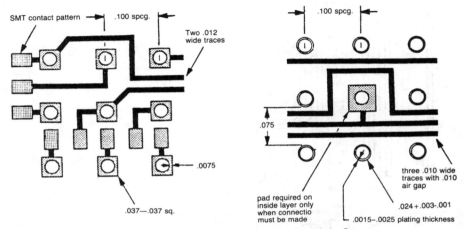

Figure 4.1 Conductor-to-via clearances. (*From ANSI/IPC-SM-782.*[3])

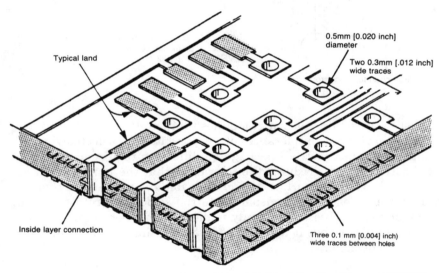

Figure 4.2 Land pattern-to-via relationships, 2.544/0.100″ grid. (*From ANSI/IPC-SM-782.*[3])

necting structure. The design of land patterns is very critical because it is the land pattern that not only determines the solder joint strength and hence the reliability of solder joints but also impacts the solder defects, cleanability, testability, and repair and rework. The very producibility or the success of the assembly is dependent upon the land pattern design.

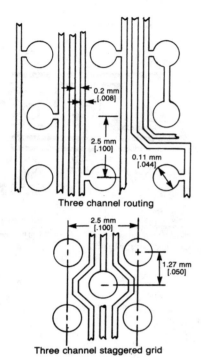

Three channel routing

Three channel staggered grid

Figure 4.3 Three conductor per channel routing concepts. (*From ANSI/IPC-SM-782.*[3])

There are certain general guidelines that one should follow to cope with the variations in tolerances of components. The selected vendor's components must pass all package qualification requirements. Standardization of parts reduces the tolerances that the land pattern design will have to support. A second desirable requirement is that the land pattern design be transparent to the soldering process used in manufacturing. This will not only reduce the number of land sizes in the design library but it will also be less confusing for the printed circuit board designer.

4.5.1 Layout scale

The design and artwork preparation of the board is the most vital element in achieving manufacturing success in SMT. Because the surface-mount components are half the size of the leaded devices, the scale of the layout must be carefully considered. On larger boards with a mix of leaded and surface-mount components, a 2:1 scale may be adequate. More complex surface mount layouts can be designed at a 4:1 scale with excellent results. The larger scale will make it possible to increase density and assure accuracy. When designing on a CAD system, the designer can gain accuracy and density by increasing the grid resolution.

4.5.2 Land geometry

Land geometry plays a prominent role in a successful assembly. If the land geometry is too long, the component may float to one side. If the pattern is too wide, the component may rotate. The ideal pattern would cool evenly and center the component.

4.5.3 Sharing lands

Designers should avoid shortcuts in designing for surface-mount devices. For example, do not attempt to use a single large land between two components. The solder on the larger land will overwhelm the smaller outter lands and draw the components to the larger deposit of solder. This may result in an excess of solder buildup on the shared land during wave soldering.

4.5.4 Land pattern separation

Separation of land patterns will ensure containment of the solder paste. If it is important to increase conductor width between components, the designer will have better results with two narrow traces rather than one wide trace. The containment of the solder by separation of the land patterns is the key to controlling the quality of the soldering process, as shown in Fig. 4.4.

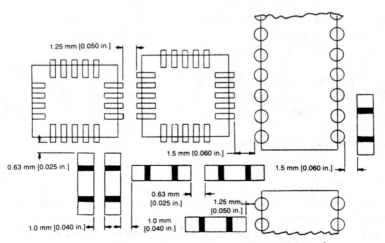

Figure 4.4   Land-to-land clearances.   (*From ANSI/IPC-SM-782.*[3])

The minimum land-to-land spacing between adjacent components should be 1.25 mm (0.050 in). For plastic leaded chip carriers this means 1.9 mm (0.075 in) between leads of adjacent components. For components where leads or terminations are only on two sides of the packages, the interpackage spacing should be 1.25 mm (0.050 in) for J-lead SOICs and 1.0 mm (0.040 in) for discrete passive components between the adjacent sides that do not have leads or terminations.

4.5.5   Conductors under components

Routing conductors under components is usually unavoidable. Usually the narrow conductors will not cause a problem, but care should be taken to allow adequate clearance for cleaning of flux and debris from under the components. However, wide conductors under a component can cause components to be displaced by solder migration during the reflow process. The component can then lift away from the contact area, causing unwanted secondary rework of the assembly.

4.5.6   Adhesive control

When surface-mount components are bonded to the substrate prior to soldering, controlling the amount of adhesive is important to ensure a good bond. Too much adhesive applied to the bonding area may result in the excess being forced out from under the part onto the lands. A minimal amount of adhesive, sufficient to provide bonding, is desirable.

To facilitate adhesive control, some land patterns have incorporated dummy lands in the bonding area so that the adhesive bond is reduced to the distance between the underside of the component and the top of the conductor, as opposed to the top of the P&I structure. This technique may be accomplished by the incorporation of a nonfunctional (dummy) land between the component mounting points or having actual conductors pass between the lands used for mounting the components. When high-aspect-ratio solder bumps are deposited on the P&I structure prior to assembly (plated tin-lead and solder paste reflowed together), the nonfunctional land used as a bonding site can significantly reduce the amount of adhesive required to make a proper bond.

4.5.7   Test points

Whenever possible, the design should incorporate test probe locations to aid not only in circuit testing but also in helping to diagnose problem areas. Location of test probes is usually a function of the electrical requirements of the circuit.

4.5.8 Solder mask clearances

A solder mask may be used to isolate the land pattern from other conductive features on the board such as vias, lands, or conductors. Where no conductors run between lands, a simple gang mask can be used. A 0.38-mm (0.015-in) spacing could be acceptable. Because of the close proximity of the solder mask to the land pattern, care must be taken in choosing a mask that has low-flow and low-solvent-bleed characteristics to avoid land pattern contamination.

Solder paste used in mounting of surface-mount components has two undesirable characteristics which can be helped by the use of a solder mask. The solder paste may leave small uncoalesced balls of solder as it melts. Secondly, the flux tends to cause the solder paste to flow out and cover adjacent conductors. Both of these problems can be helped by a solder mask that forms a complete pocket around each land. The pocket tends to prevent the flow of the paste onto adjacent conductors and into holes and reduces the possibility of solder ball shorts.

4.6 Land Pattern Details

ANSI/IPC-SM-782 provides information on land pattern geometries used for the surface attachment of electronic components. The intent of the information it presents is to provide the appropriate size, shape, and tolerance of surface-mount land patterns to ensure sufficient area for the appropriate solder fillet, and to also allow for inspection and testing of those solder joints. Although in many instances the land pattern geometries can be slightly different based on the type of soldering used to attach the electronic part, the information provided in ANSI/IPC-SM-782 is intended, wherever possible, to define a pattern in such a manner that it is transparent to the attachment process being used.

Designers should be able to use the information contained therein to establish standard configurations not only for manual designs but also for CAD systems. Whether parts are mounted on one or both sides of the board, subjected to wave, reflow, or other type of soldering, the land pattern and part dimensions should be optimized to ensure proper solder joint and inspection criteria.

The following criteria and illustrations are examples of the information provided in ANSI/IPS-SM-782 for components described in Electronics Industry Association documents EIA-IS-30-A (resistors)[9], EIA-CB-11 (capacitors)[10], and JEDEC-95 (solid-state products).[11]

All land patterns are detailed using nominal dimensions for the circuit land configuration. The nominal dimensions are shown locating the center of each land, related to each other. Reference dimensions are shown to provide inspection criteria which must take into consideration the manufacturing tolerances associated with circuit definition on the mounting substrate. All land patterns are provided making maximum use of a 0.5-mm (0.020-in) grid for the land pattern layout. Land patterns for each part are described in terms of the placement outline needed to allow electrical clearance of all the lands located within the part shape and land configuration outline. Placement real estate is expressed in terms of the grid elements [0.5 mm (0.020 in)] needed to accommodate the part and its land patterns.

4.6.1 Basic criteria

A common land pattern design guideline for both reflow- and wave-soldered assemblies for resistors is based on the following formulas:

$$\text{Land width } (X) = W_{max} - K$$

$$\text{Land length } (Y) = H_{max} + K$$

$$\text{Gap between lands } (A) = L_{max} - 2T_{max} - K$$

where W = width of component
H = component height
L = component length
T = the solderable termination
K = a constant of 0.25 mm (0.010 in)

All final numbers derived from this equation should be rounded to the nearest sensible land size, taking into account that board-processing tolerances can reduce or enlarge the conductive pattern on the P&I structure by some amount depending on the class.

For capacitors the formula is modified slightly. Determining land width X and gap A is the same; however, the land length changes due to the height of the metallization. The land length formula should be:

$$Y = H_{max} + T_{min} - K$$

The equations may be used to determine the land geometry for any surface-mount part. Careful analysis of the solder joint junction should be accomplished to ascertain that a solder pedestal can be achieved.

4.6.2 Discrete rectangular components

Land patterns for discrete rectangular components (resistors and capacitors) are designed to provide the mounting surface for an appropriate solder pedestal. The concepts detailed in Sec. 4.6.1 have been

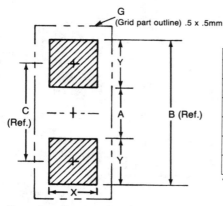

Code Letter	RC 0805	RC 1206*	RC 1210
A	0.8 [.032]	1.8 [.070]	1.8 [.070]
B	3.8 [.150]	5.0 [.200]	5.0 [.200]
C	2.3 [.090]	3.4 [.134]	3.4 [.134]
X	1.4 [.055]	1.6 [.063]	2.6 [.102]
Y	1.5 [.060]	1.6 [.063]	1.6 [.063]
Grid part outline	4 x 8	4 x 12	6 x 12

* Example shown in graphics

Figure 4.5 Rectangular chip resistor land pattern. (*From ANSI/IPC-SM-782.*[3])

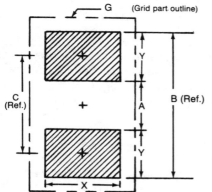

Code Letter	CC 0805	CC 1206	CC 1210*	CC 1812	CC 1825
A	0.8 [.032]	1.8 [.070]	1.8 [.070]	3.2 [.126]	3.2 [.126]
B	3.8 [.150]	5.0 [.200]	5.4 [.213]	6.8 [.268]	6.8 [.268]
C	2.3 [.090]	3.4 [.134]	3.6 [.142]	5.0 [.200]	5.0 [.200]
X	1.4 [.055]	1.6 [.063]	2.6 [.102]	3.2 [.126]	6.6 [.260]
Y	1.5 [.060]	1.6 [.063]	1.8 [.070]	1.8 [.070]	1.8 [.070]
Grid part outline	4 x 8	4 x 12	6 x 12	8 x 16	14 x 16

* Example shown in graphics

Figure 4.6 Rectangular ceramic chip capacitor land pattern. (*From ANSI/IPC-SM-782.*[3])

applied to the evaluation of land pattern geometries for the rectangular components. Figure 4.5 provides the appropriate land pattern for chip resistors.

The grid part outline may be used to place these parts in a configuration which the outlines are touching; under these conditions there is sufficient electrical clearance between adjacent land patterns and mechanical clearance if the parts are properly mounted. In addition, Fig. 4.6 provides the land patterns for the ceramic capacitors.

4.6.3 Tubular components

The land pattern concepts that are described in Sec. 4.6.1 may also be used to develop the land pattern geometries for discrete tubular com-

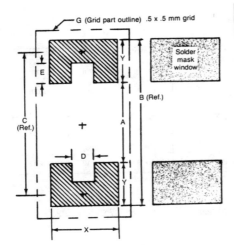

Code Letter	MELF 1/4 W	MLL34	SOD80	MLL41*
A	4.4 [.174]	2.2 [.087]	2.2 [.087]	3.4 [.134]
B	8.4 [.330]	5.0 [.200]	5.4 [.213]	7.0 [.276]
C	6.4 [.252]	3.6 [.142]	3.8 [.150]	5.2 [.205]
D†	0.3 [.012]	0.3 [.012]	0.3 [.012]	0.3 [.012]
E†	0.8 [.032]	0.7 [.028]	0.8 [.032]	1.0 [.040]
X	2.5 [.100]	2.0 [.080]	2.0 [.080]	2.8 [.110]
Y	2.0 [.080]	1.4 [.055]	1.6 [.063]	1.8 [.070]
G	6 x 18	6 x 12	6 x 12	6 x 16

* Example shown in graphics

† The notch shown by dimensions "D" and "E" are intended to reduce skewing and are optional. The depth of this notch may be determined by the following:

$$E = Y - \left(\frac{B - L_{MAX}}{2}\right)$$

Where "L" equals the maximum body length.

Figure 4.7 Tubular discrete component land pattern. (*From ANSI/IPC-SM-782.*[3])

ponents. However, cutouts in rectangular areas have been found to aid holding the component in place during an automatic reflow process. They may be omitted if other processes are used. Figure 4.7 provides the nominal dimensions for land patterns used to mount tubular components.

4.6.4 Small-outline transistors

Recommended land patterns for commonly used small outline transistors (SOTs) are shown in Fig. 4.8 (SOT-23), 4.9 (SOT-89), and 4.10 (SOT-143).

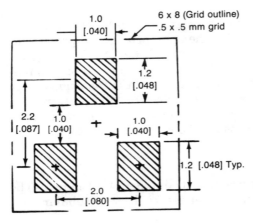

Figure 4.8 SOT-23 transistor land pattern. (*From ANSI/IPC-SM-782.*[3])

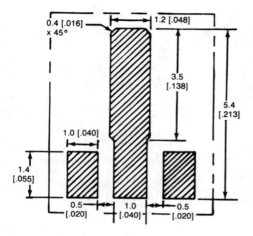

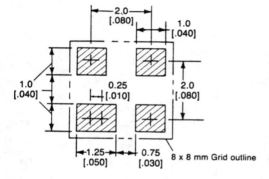

Figure 4.9 SOT-89 transistor/ diode land pattern. (*From ANSI/IPC-SM-782.*[3])

Figure 4.10 SOT-143 diode land pattern. (*From ANSI/IPC-SM-782.*[3])

4.6.5 Gull-wing integrated circuits

The gull-wing (L-shaped) leads of SOIC packages are prone to damage during handling. There is also great variation between the heels of SOIC dimensions. Because of this great variation between the heels of SOIC dimensions. Because of this great variation in dimension between the heels of these components, it is better to use the body width than the gap between the heels as the reference dimension in formulating land pattern designs of SOICs. The recommended land pattern guidelines for an SOIC are as follows:

Land width = 0.63 mm (0.025 in)

Land length = 2.0 mm (0.080 in)

A = maximum component body dimension − K

where K = constant; it is recommended to be 0.25 mm (0.010 in). Us-

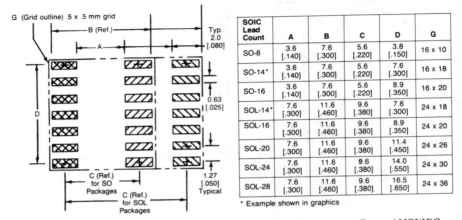

SOIC Lead Count	A	B	C	D	G
SO-8	3.6 [.140]	7.6 [.300]	5.6 [.220]	3.8 [.150]	16 x 10
SO-14*	3.6 [.140]	7.6 [.300]	5.6 [.220]	7.6 [.300]	16 x 18
SO-16	3.6 [.140]	7.6 [.300]	5.6 [.220]	8.9 [.350]	16 x 20
SOL-14*	7.6 [.300]	11.6 [.460]	9.6 [.380]	7.6 [.300]	24 x 18
SOL-16	7.6 [.300]	11.6 [.460]	9.6 [.380]	8.9 [.350]	24 x 20
SOL-20	7.6 [.300]	11.6 [.460]	9.6 [.380]	11.4 [.450]	24 x 26
SOL-24	7.6 [.300]	11.6 [.460]	9.6 [.380]	14.0 [.550]	24 x 30
SOL-28	7.6 [.300]	11.6 [.460]	9.6 [.380]	16.5 [.650]	24 x 36

* Example shown in graphics

Figure 4.11 Small-outline integrated circuit (SOIC) land pattern. (*From ANSI/IPC-SM-782.*[3])

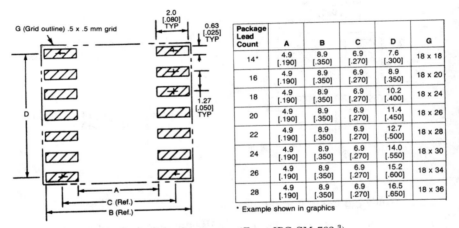

Package Lead Count	A	B	C	D	G
14*	4.9 [.190]	8.9 [.350]	6.9 [.270]	7.6 [.300]	18 x 18
16	4.9 [.190]	8.9 [.350]	6.9 [.270]	8.9 [.350]	18 x 20
18	4.9 [.190]	8.9 [.350]	6.9 [.270]	10.2 [.400]	18 x 24
20	4.9 [.190]	8.9 [.350]	6.9 [.270]	11.4 [.450]	18 x 26
22	4.9 [.190]	8.9 [.350]	6.9 [.270]	12.7 [.500]	18 x 28
24	4.9 [.190]	8.9 [.350]	6.9 [.270]	14.0 [.550]	18 x 30
26	4.9 [.190]	8.9 [.350]	6.9 [.270]	15.2 [.600]	18 x 34
28	4.9 [.190]	8.9 [.350]	6.9 [.270]	16.5 [.650]	18 x 36

* Example shown in graphics

Figure 4.12 Gullwing flatpack land pattern. (*From IPC-SM-782.*[3])

ing this formula for gull-wing component land pattern design results in the configurations shown in Fig. 4.11 and 4.12.

4.6.6 Plastic leaded chip carriers

The land pattern design guidelines for plastic leaded chip carriers are based on the following formulas:

Land width = 0.63 mm (0.025 in)

Land length = 2.0 mm (0.080 in)

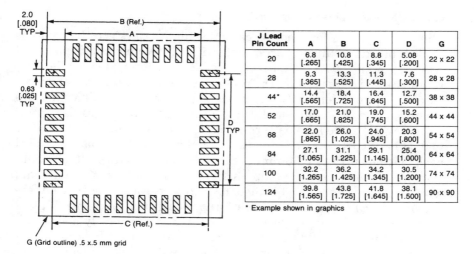

J Lead Pin Count	A	B	C	D	G
20	6.8 [.265]	10.8 [.425]	8.8 [.345]	5.08 [.200]	22 x 22
28	9.3 [.365]	13.3 [.525]	11.3 [.445]	7.6 [.300]	28 x 28
44*	14.4 [.565]	18.4 [.725]	16.4 [.645]	12.7 [.500]	38 x 38
52	17.0 [.665]	21.0 [.825]	19.0 [.745]	15.2 [.600]	44 x 44
68	22.0 [.865]	26.0 [1.025]	24.0 [.945]	20.3 [.800]	54 x 54
84	27.1 [1.065]	31.1 [1.225]	29.1 [1.145]	25.4 [1.000]	64 x 64
100	32.2 [1.265]	36.2 [1.425]	34.2 [1.345]	30.5 [1.200]	74 x 74
124	39.8 [1.565]	43.8 [1.725]	41.8 [1.645]	38.1 [1.500]	90 x 90

* Example shown in graphics

G (Grid outline) .5 x.5 mm grid

Figure 4.13 J-lead postmolded (plastic) square-outline chip carrier land pattern. (*From ANSI/IPC-SM-782.*[3])

$$B = \text{Maximum component body dimension} + K$$

where K = constant; it is recommended to be 0.75 mm (0.030 in). Based on the above formulas, the land pattern design for plastic leaded chip carrier packages is summarized in Fig. 4.13 for the square-outline J-leaded chip carrier and Fig. 4.14 for the rectangular-outline J-leaded chip carrier.

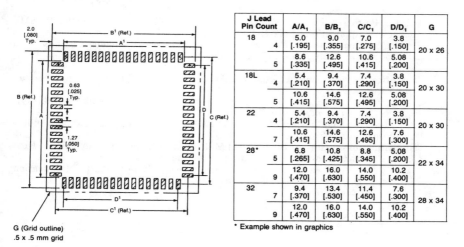

J Lead Pin Count		A/A_1	B/B_1	C/C_1	D/D_1	G
18	4	5.0 [.195]	9.0 [.355]	7.0 [.275]	3.8 [.150]	20 x 26
	5	8.6 [.335]	12.6 [.495]	10.6 [.415]	5.08 [.200]	
18L	4	5.4 [.210]	9.4 [.370]	7.4 [.290]	3.8 [.150]	20 x 30
	5	10.6 [.415]	14.6 [.575]	12.6 [.495]	5.08 [.200]	
22	4	5.4 [.210]	9.4 [.370]	7.4 [.290]	3.8 [.150]	20 x 30
	7	10.6 [.415]	14.6 [.575]	12.6 [.495]	7.6 [.300]	
28*	5	6.8 [.265]	10.8 [.425]	8.8 [.345]	5.08 [.200]	22 x 34
	9	12.0 [.470]	16.0 [.630]	14.0 [.550]	10.2 [.400]	
32	7	9.4 [.370]	13.4 [.530]	11.4 [.450]	7.6 [.300]	28 x 34
	9	12.0 [.470]	16.0 [.630]	14.0 [.550]	10.2 [.400]	

* Example shown in graphics

G (Grid outline) .5 x .5 mm grid

Figure 4.14 J-lead postmolded (plastic) rectangular-outline chip carrier land pattern. (*From ANSI/IPC-SM-782.*[3])

4.6.7 Ceramic leadless chip carriers

The land patterns for leadless ceramic chip carriers are the same as leaded plastic counter parts, but with a recommended value of $K = 1.75$ mm (0.070 in). The value of K is different for leadless ceramic chip carriers because the package sizes for the ceramic and plastic devices are different. Ceramic packages, for a given pin count, are about 1.0 mm (0.040 in) smaller than the plastic packages.

Having the same land pattern for both types of packages allows 0.25 to 0.38 mm (0.01 to 0.015 in) outside land extension for plastic leaded packages but 0.75 to 0.88 mm (0.030 to 0.035 in) outside land extension for leadless ceramic chip carriers. The value of $K = 1.75$ mm (0.070 in) is recommended if the substrate's coefficient of thermal expansion (CTE) matches closely with that of ceramic packages. If glass epoxy substrate is used (glass epoxy substrate is generally not recommended) then K should be 2.25 mm (0.090 in) to give further land extension and larger land design, 2.5 mm (0.100 in) should be used.

References

1. Vern Solberg, NuGrafix Group, "Design Guidelines for SMT," *Printed Circuit Design*, January 1986.
2. Brian T. Patterson, Kollmorgen Corp., "Tips for PWB Design, Construction for SMT Technology Application—Part 1," *Surface Mount Technology*, April 1986.
3. "Component Packaging and Interconnecting with Emphasis on Surface Mounting," ANSI/IPC-SM-782, July 1988, Institute for Interconnecting and Packaging Electronic Circuits, Lincolnwood, Ill.
4. "Guidelines for Selecting Printed Wiring Board Sizes Using Standard Panel Sizes," ANSI/IPC-D-322, September 1990, Institute for Interconnecting and Packaging Electronic Circuits, Lincolnwood, Ill.
5. Brian T. Patterson, Kollmorgen Corp., "Tips for PWB Design, Construction for SMT Technology Application—Part 2," *Surface Mount Technology*, June 1986.
6. Steven W. Hinch, Hewlett-Packard Co., "Surface Mount Technology,": Chap. 3, in C.F. Coombs, Jr. (ed.), *Printed Circuits Handbook*, 3d ed., McGraw-Hill, New York, 1988.
7. "Signetics Surface Mount Process and Application Notes," 1989, North American Philips Corporation.
8. Ray P. Prasad, Intel Corp., "Designing Surface Mount for Manufacturability," *Printed Circuit Design*, May 1987.
9. "Resistors, Rectangular Surface Mount Thick Film," EIA-IS-30-A, Electronic Industries Association, Washington, D.C.
10. "Guidelines for the Surface Mounting of Multilayer Ceramic Chip Capacitors," EIA-CB-11, Electronic Industries Association, Washington, D.C.
11. "JEDEC Registered and Standard Outlines for Solid State Products," Publication 95, Electronic Industries Association, Washington, D.C.

Design for Automatic Through-Hole Assembly*

5.1 Introduction[1]

The design criteria described in this chapter establish the considerations for maximizing machine insertion of leaded electronic components into printed circuit boards. It covers axial, radial, and DIP devices, primarily, although it also gives references for certain other "leaded" components.

5.2 General Considerations

In the application of machine insertion to printed circuit board assembly, the following general rules should be applied. Each of these is described in subsequent paragraphs of this chapter.

- Standardize tooling for printed circuit board mounting.
- Utilize known references.
- Orient components properly.
- Select standardized components whenever possible.
- Select proper printed circuit board hole diameters.
- Provide proper clearances between components.
- Provide proper clearances between tooling holes and components.
- Provide proper clearances for insertion head and cut and cinch.

5.2.1 Printed circuit board size

The maximum size of the printed circuit board is generally related to the insertable area of the insertion machine used and/or by an auto-

---

*This chapter is based exclusively on information provided by Universal Instruments Corporation Binghamton, N.Y.

matic board-handling system. Standard insertable areas vary according to the machine selected. Most manually fed, processor-controlled machines usually have a standard insertable area of 457.2 mm by 457.2 mm (18 × 18 in).

Where small printed circuit boards are required, it is sometimes advantageous to use a "breakaway" or "biscuit" board. This configuration consists of a series of small boards which are punched or drilled within a single larger board. Once inserted, they can be separated on perforated or grooved edges, or they may be routed or sheared. A technique called "press-back" may also be used when small boards are punched out of a larger sheet. After punching the outline, it is pressed back into the larger board and held by friction until the assembly and soldering process is complete. Breakaway boards, or boards sheared after assembly, minimize board handling throughout all phases of board production, particularly when only a few components are inserted into each printed circuit board.

5.2.2 Locating references

Of prime importance in printed circuit board design and construction is the establishment of accurate reference points, or locating references, to which all holes are drilled or punched. Locating references generally take the form of two holes. Other possible approaches are "V" notches, multiple slots, or straight surfaces. These references are used to position the printed circuit board on the workboard holder for automatic insertion. It is recommended that the distance between the locating references be the maximum separation permitted by the length of the board. Also, it is recommended that the locating reference hole diameter tolerance be held to ± 0.025 mm (0.001 in). The lower left reference hole is generally given the location terminology of X_o, Y_o.

5.2.3 Insertion hole locations

The locations of all insertion holes should be defined relative to the board-locating references. The accuracy of the hole locations with respect to the locating references will affect insertion reliability and hole diameter requirements. With drilled boards, the hole tolerance from the board references should be ± 0.076 mm (0.003 in) or less. With punched boards, the reference holes or surfaces should be punched using the same die set used for the insertion holes and not a part of the blanking die.

Particularly for boards with drilled holes, care must be exercised in the initial board layout to ensure that the insertion hole locations and the board locating references are only "one tolerance" apart. If this consideration is not met, tolerance buildup during insertion will reduce insertion reliability. The "one tolerance" requirement can easily be accomplished by establishing X axis and Y axis dimensional reference lines through the centerlines of the locating references and using these predrilled holes to locate the board for insertion hole drilling.

If both the reference holes and insertion holes are drilled at the same setup, the net accuracy is two times the normal drilling machine tolerance. Once these lines are established, they may be used as the references for drilling or punching of the printed circuit board, as well as for insertion.

5.2.4 Component location objectives

5.2.4.1 Axis considerations. The most efficient use of automatic insertion equipment (Fig. 5.1), and the greatest component population density can be achieved by inserting components in only one or two orthogonal axes. The use of only one axis minimizes the amount of board handling and machine operations.

Two-axes insertion is an acceptable and efficient way of inserting

Figure 5.1 Two-axes insertion. *(Courtesy of Universal Instruments.)*

components. There are two common methods of performing two-axes insertion. One method, used on earlier machines with a rotary table, involves designing the workboard holder, which supports the printed circuit board on the machine table, such that one window (printed circuit board location) supports the printed circuit board in an X axis while another window supports it in the Y axis. The operator must load each printed circuit board on the workboard holder twice.

Using the second method, a properly designed workboard holder is rotated in 90° increments. Using the rotary table, now standard, rotation can be automatically accomplished as part of the insertion process. However, one must recognize that the use of the rotary table contributes to a tolerance buildup which can lower insertion reliability. With automatic board handling, a rotary table is required for two-axes insertion. Applications which use more than two axes are generally inefficient and difficult to implement.

5.2.4.2 Component selection. Some operations allow a wide range of components to be inserted with minimum tooling changes. Careful component selection, however, can provide improved automatic insertion reliability. It is wise, whenever possible, to limit the range of component body sizes and lead diameters used. Doing this will decrease the need for editing of established programs and generally reduce the need for time-consuming tooling changes.

Certain systems offer a selection of programmed component lead spans and body diameters. This technology has been developed to its fullest capability in today's processor-controlled systems to allow the needed flexibility to use components that today are readily available throughout the world.

5.2.4.3 Lead hole diameter considerations. Each general type of component insertion has its own rule for establishing hole diameter. Hole diameter is basically a function of the machine board accuracy and lead diameter. In each section of this chapter there is a paragraph on hole diameter considerations for the specific type of component involved. For higher yield rates, it must be noted that component tolerances, board tolerances, and hole diameters play an important role. Pushing any or all of these to their limits will lower yields.

While some hole diameters may seem somewhat large in size, they take into account typical conditions on the production floor. They deal with nominal dimensions, plus the tolerances of all elements of the insertion system regardless of type. It is one thing to say components can be inserted into smaller holes, and it is another thing to accomplish it in a production environment. Taking this into account, the hole sizes established in these guidelines can generally be used without special considerations and with excellent results for production yield.

There have been occasions where customers have successfully inserted components into holes smaller than those recommended. Usually there are some other compensating factors involved, such as very small board size, better hole location tolerance, or a recognized lower yield rate.

5.2.4.4 Location considerations. Automatic insertion of printed circuit boards requires that there be sufficient clearances around components and locating references to allow room for the insertion head and cut and clinch tooling. The clearances required will vary with the type of insertion machine.

5.2.5 Board holder design

It is necessary to firmly retain and support the printed circuit board on the insertion machine during insertion, and to provide an easy method of board loading and unloading. This is done by means of a workboard holder, a metal plate with specially designed openings and accurate locating means. When designing a workboard holder, the following general rules should be applied:

- Support the printed circuit board as much as possible.
- Provide finger room for loading and unloading for manually loaded machines.
- Utilize known references.
- Provide proper clearances around locating references.
- Provide board hold-down during insertion.

Blank workboard holders are usually available from the machine vendor. Typically these are precision, flat ground aluminum, approximately 6.35 mm (0.25 in) thick, containing precision locating holes with steel bushings placed in the four corners. There are also holes provided for hold-down screws to mount the workboard holder to the machine.

5.2.6 Workboard holder considerations

5.2.6.1 Board openings. Board openings, or windows, must be positioned so that all components to be inserted are located within the insertable area of the insertion machine. Maximum insertable area begins at machine zero, located a defined distance from the lower left table reference bushing, and proceeds in a positive X and Y direction from that point.

When designing workboard holder openings, it is desirable to support the printed circuit board as much as possible with at least 3.18 mm (0.125 in) of workboard holder surface extending beneath the board at all support areas. Supporting the board at all four corners is preferable; larger boards may require additional support near the middle of long sides.

A certain amount of edge clearance is required for cut and clinch clearance when inserting components close to the edge of the board. Where manual board loading and unloading is used, additional finger room should be provided at the right and left sides of the windows to allow easy loading and unloading of the printed circuit board.

The acceptable distance between the edge of the board opening and an insertion varies with the type of component, machine, insertion head, and cut and clinch unit. In each section of this chapter, there is a paragraph on uninsertable area for the specific type of component involved. Refer to the section that applies to the specific component type to be inserted to determine its uninsertable area. In general, printed circuit designs requiring components inserted close to locating references and board edges may result in loss of automatic insertion of these components due to board support requirements.

5.2.6.2 Locating methods. Locating holes, V notches, slots, or other accurate surfaces may be used for locating the printed circuit board on the workboard holder for insertion. Regardless of the method used, it is required that two locations be utilized, and that the distance between them be the maximum permitted by the length of the board. The same locating references from which all the holes are drilled or punched should be employed for this purpose.

Pins through two locating holes are the most common and preferred locating method; they may be mounted directly into the workboard holder or into inserts which are mounted onto the workboard holder. Inserts allow insertions to be made closer to the locating references than do workboard holder–mounted locating techniques, as well as providing necessary support in a relatively limited support area.

5.2.6.3 Hold-down devices. Hold-down devices are used to maintain board position during table movement and component insertion. The most common of these are clamp assemblies which must be manually positioned during board load and unload or spring stops which use spring loaded plungers to apply pressure to the edges of the board to hold it in position. Hold-down devices are mounted at any place where the board is supported and, most advantageously, at opposite sides of the printed circuit board near the locating pins.

5.2.7   Rotary workboard holder design

Rotary workboard holders usually have one, two, or four board openings. These locations are usually arranged in a symmetrical pattern about the center of the workboard holder to allow component insertion in both 0 and 90° axis positions. For manually loaded machines, finger room for board loading and unloading should be provided in both the X axis and Y axis if the workboard holder is to be rotated.

5.2.8   Pass-through workboard holders

Two types of board location methods are often available, i.e., front-edge-justified or centerline-justified. Front-edge-justified means the front edge of the board is fixed and the board "grows" to the rear of the machine. Centerline-justified means that the centerline of the board on the table is fixed and everything "grows" symmetrically. Either system may be ordered, however, all machines in the system must use the same method. The major advantage of the front-edge-justification approach is that a standard, adjustable workboard holder may be used. If centerline justification is used, the workboard holder becomes special. In general, the clearances for support and locating are similar, but because of the fixed front and rear edge support and the automatic locating and hold-down means, the uninsertable areas are usually greater.

Some machines utilize the standard rotary table. From a board design standpoint, edge clearances on certain components, primarily DIPs may be placed as close to the rear edge as you can get to the front edge with zero rotation. Again, footprint considerations and precedence are of extreme importance. For wide boards, e.g., 20 cm (8 in) or more, and/or for boards which have been weakened or for those which will have a heavy component load by the end of the last automatic inserter, some form of board support at or near the centerline may be required to control board "sag."

Precision positioning of a printed circuit board on the workboard holder is usually accomplished by locator holes. In addition, the width of the board W and the length L must be maintained. This is to ensure proper transfer of the board from the magazine (or stack) through the machine and into the output magazine as edge support is used throughout. This also limits the insertable area.

5.3 Programming Considerations

Processor-controlled machines require that a pattern program be entered into the controller memory. The pattern program contains all the information required to automatically populate a printed circuit board. In most cases, each printed circuit board to be inserted or populated is considered one complete pattern. As a result, it is necessary to define the insertion locations only once, regardless of the number of boards that are mounted on the workboard holder, and also in the case of breakaway boards. All insertion locations are defined relative to one point on the board, hereafter referred to as the "pattern program zero" reference.

The pattern offsets then define the distance between the machine zero and the printed circuit board datum reference. One offset is entered for each printed circuit board location (window) on the workboard holder and/or each printed circuit board on a breakaway board. The offsets for a typical printed circuit board are shown in Fig. 5.2.

The generation of the pattern program should take into consideration many things in addition to the above basics to produce optimum results. Some of these other considerations are:

- *Path:* A minimum travel path to cover the board in the least total time. There are suggestions to optimize this in each chapter.

- *Clearances:* Including both head, anvil and workboard holder clearances.

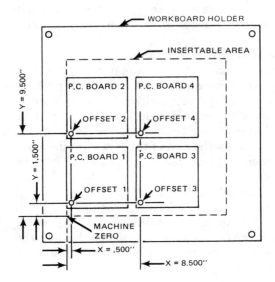

Figure 5.2   Pattern offsets.

- *Precedence:*   Which component should be programmed first? Generally, the least expensive are often done first to minimize possible scrap costs; however, sometimes size requires that certain components be done last like radials and square-wire pins. Other considerations are ease of repair and tooling footprint.

- *Magazine selection versus speed:*   On DIP inserters, where shuttle travel is a factor in establishing insertion rate, component values used most frequently should be located closest to the insertion head.

5.4   Axial-Lead Component Insertion

Equipment manufacturers offer a wide variety of machines for the automatic and semi-automatic insertion of axial lead components, i.e., from manually loaded, single-head, variable-center distance (VCD) inserters through dual-head VCDs, plus the ability to incorporate fully automatic board handling on single-head pass-through versions (Fig. 5.3). Several machines are equipped with rotary tables and some machines use microprocessor control. This means that they can automatically insert a wide range of presequenced component types.

Figure 5.3   Single-head axial-lead inserter.   *(Courtesy of Universal Instruments.)*

5.4.1 Component Input Taping Considerations[2]

Component taping requirements for most axial-lead processing equipment generally comply with Electronics Industry Association (EIA) Standard RS-296-D, which is accepted as the industry standard. When used as the input to a VCD, the sequencer output must meet the VCD-head input specifications as well. If nonsequenced taped components are to be used in a VCD, they must also meet this same specification.

5.4.2 Axial-lead sequencing

Sequencing, which in this case is the preparation of axial-leaded components prior to insertion on a VCD, usually permits complete or nearly complete insertion of this type of a component on a given printed circuit board. With a typical sequence, the components are cut from the input tape, dispensed onto a moving conveyor, and fed into a special taping mechanism where they are accurately centered, repitched, and retaped to a distance between tapes somewhat smaller than the original input reel. As a consequence, sequenced components do not use the same maximum insertion spans obtained with standard taped components.

Once the input class has been tentatively selected, the next task is to check the insertion head specifications to see if this class will allow the desired insertion spans. It is not generally recommended that class changes be planned as part of an operating routine because too many items must be reset and readjusted.

5.4.3 Insertion center considerations

Axial-lead insertion heads are sometimes similar in design and function. Figure 5.4 shows the input and output characteristics of a typical VCD insertion head. Minimum VCD spans are further limited by component body length. To be more precise, one may use the following:

Minimum insertion span = body length + 0.062 in

+ (inside former width × 2) + one wire diameter

or, for metric,

Minimum insertion span = Body length + 1.57 mm

+ (inside former width × 2) + one wire diameter

In the above formulas, the 1.57-mm (0.062-in) value accounts for component centering variations, which is two times 0.79 mm (0.031 in) (0.79 mm) (Fig. 5.4, Note 1), and the "one" wire diameter used is re-

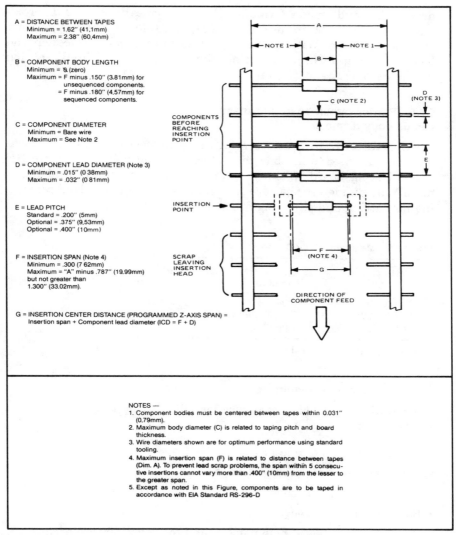

A = DISTANCE BETWEEN TAPES
  Minimum = 1.62" (41.1mm)
  Maximum = 2.38" (60.4mm)

B = COMPONENT BODY LENGTH
  Minimum = 8 (zero)
  Maximum = F minus .150" (3.81mm) for
       unsequenced components.
     = F minus .180" (4.57mm) for
       sequenced components.

C = COMPONENT DIAMETER
  Minimum = Bare wire
  Maximum = See Note 2

D = COMPONENT LEAD DIAMETER (Note 3)
  Minimum = .015" (0.38mm)
  Maximum = .032" (0.81mm)

E = LEAD PITCH
  Standard = .200" (5mm)
  Optional = .375" (9.53mm)
  Optional = .400" (10mm)

F = INSERTION SPAN (Note 4)
  Minimum = .300 (7.62mm)
  Maximum = "A" minus .787" (19.99mm)
  but not greater than
  1.300" (33.02mm).

G = INSERTION CENTER DISTANCE (PROGRAMMED Z-AXIS SPAN) =
  Insertion span + Component lead diameter (ICD = F + D)

NOTES —
1. Component bodies must be centered between tapes within 0.031"
   (0.79mm).
2. Maximum body diameter (C) is related to taping pitch and board
   thickness.
3. Wire diameters shown are for optimum performance using standard
   tooling.
4. Maximum insertion span (F) is related to distance between tapes
   (Dim. A). To prevent lead scrap problems, the span within 5 consecu-
   tive insertions cannot vary more than .400" (10mm) from the lesser to
   the greater span.
5. Except as noted in this Figure, components are to be taped in
   accordance with EIA Standard RS-296-D

Figure 5.4 Typical input specifications—VCD insertion head.

ally the two inside halves. The typical width of the inside former is
0.81 mm (0.032 in). Note that the programmed Z axis span is the selected insertion span plus one wire diameter (the two outside halves).

5.4.4 Lead form and tooling

The VCD head is generally tooled for the right-angle lead form. The
end of component body to inside of form in lead is a function of the

driver tips and inside formers selected. Typically this is approximately 2 mm (0.080 in) minimum. Larger values of X will give more reliable insertion. With special attention to component selection and body centering during sequencing, the 1.91 mm (0.075 in) dimension can be used as the minimum.

5.4.5   Printed circuit board thickness versus body diameter

Another factor to be considered for the VCD head is the board thickness versus maximum body diameter, since overall cutoff length is fixed. Taking board thickness into consideration, the typical maximum body diameters that are insertable with sufficient lead remaining for cutting and clinching are shown in Table 5.1 and Fig. 5.5.

TABLE 5.1   Maximum Body Diameter

Board thickness, mm (in)	Standard tooling, mm (in)	Optional tooling, mm (in)
2.36 (0.093)	5.94 (0.234)	9.0 (0.354)
1.57 (0.062)	7.52 (0.296)	10* (0.394)
0.81 (0.032)	9.04 (0.356)	10* (0.394)

*Maximum diameter limited by tape pitch.

5.4.6   Hole diameter requirements

In general, the hole size necessary for reliable component insertion is a function of

- Component lead diameter
- Machine insertion head tolerances
- Table positioning accuracy
- Board holder accuracy
- Printed circuit board hole pattern and tooling reference accuracy
- Component lead length desired below the board

For most axial-lead insertion machines with rotary tables, the maximum machine tolerances for the head, X-Y table, rotary table, and workboard holder are a true position accuracy of 0.12 mm (0.005 in). This value requires a starting hole diameter of 0.25 mm (0.010 in) larger than the lead diameter being inserted. The manufacturing tolerance of the printed circuit board insertion hole pattern must be added to this hole diameter.

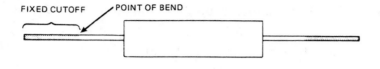

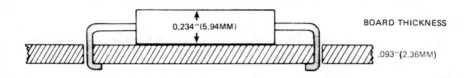

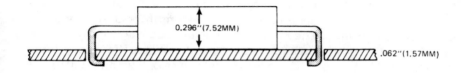

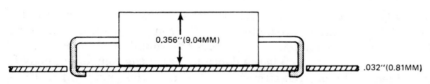

Figure 5.5 Board thickness versus axial-lead component body diameter. *(Courtesy of Standard Tooling.)*

The formula for determining hole diameter is:

Minimum hole diameter = lead diameter

+ 0.010 in + hole location tolerance

or, for metric,

Minimum hole diameter = lead diameter + 0.5 *mm*

+ hole location tolerance

For example, with a 0.25-W resistor with 0.64-mm (0.025-in) lead diameter, boards with hole location accuracy of with 0.08-mm (0.003 in) of true position, and the standard machine tolerance of 0.25 mm (0.010 in):

$$\text{Minimum hole diameter} = 0.025 \text{ in} + 0.010 \text{ in} + 0.006 \text{ in}$$

$$= 0.041 \text{ in}$$

or, for metric:

$$\text{Minimum hole diameter} = 0.64 \text{ mm} + 0.25 \text{ mm} + 0.25 \text{ mm}$$

$$= 1.04 \text{ mm}$$

These formulas normally produce insertion reliability of 99.9 percent. Reduced clearance from wire diameter to hole diameter to as low as 0.13 mm (0.005 in), may be used with small high-accuracy boards and still maintain reasonably successful insertion. However, the insertion reliability factor will be reduced. Conversely, an even higher reliability factor can be achieved by increasing hole diameter by 0.026 or 0.052 cm (0.001 or 0.002 in), by better control of sequencer centering, by increasing the span related to body length, and also by using more accurate boards. The maximum hole diameter is usually limited by soldering or good clinching requirements. When the rotary table feature is not being used, the machine tolerance of 0.25 mm (0.010 in), above, is sometimes safely reduced to 0.20 mm (0.008 in).

5.4.7   Component body configuration

Normally body configuration is not critical unless the extreme limitations of the equipment are being used. The components are handled by the leads so that component bodies may be round, oval, dog-boned in shape, etc. The *depth stop* is the depth to which the component body is assembled to the printed circuit board. Due to the lead form requirements of axial-lead components, shown in Fig. 5.6, the depth stop should be selected so that the component is held securely to the board without deforming the lead.

The VCD insertion head usually has an electrically driven depth stop cam with several program-selected letter positions in 0.20-mm (0.008 in) increments in the modifier field. With some newer single-head programmed machines, the depth stop cam can be programmed in increments of 0.25 mm (0.001 in). This value can often be modified in either direction to get the desired results. A slight looseness is recommended to avoid residual internal stresses after soldering.

5.4.8   Location considerations

5.4.8.1   Above the board. Some axial-lead insertion heads are equipped with outside formers which guide the leads to the point of

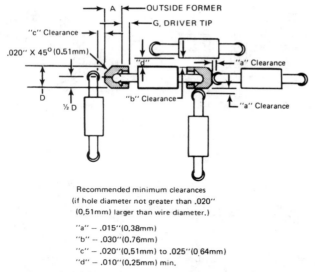

Recommended minimum clearances
(if hole diameter not greater than .020″
(0,51mm) larger than wire diameter.)

"a″ — .015″(0.38mm)
"b″ — .030″(0.76mm)
"c″ — .020″(0,51mm) to .025″(0 64mm)
"d″ — .010″(0.25mm) min.

Figure 5.6   Axial-lead component insertion tooling clearances.

insertion on the printed circuit board. Clearance around a given hole must be taken into consideration to allow the equipment to function properly. Figure 5.6 shows the top view of a typical outside former at the point of insertion, illustrating the clearances required between the lead being inserted and any adjacent component body or lead.

When the VCD insertion head employs an outside former with a single rounded V groove, the centerline of lead to former clearance line varies as a function of lead diameter. Figure 5.6 shows the largest wire diameter being inserted with conventional tooling.

For applications involving dense component assembly of small components, such as diodes or resistors of 0.25-W and smaller, outside formers are often available with smaller footprint dimensions. This, however, results in a minor loss of durability of the tool.

5.4.8.2   Below the board.   The cut and clinch is normally positioned approximately 15.2 mm (0.600 in) below the board to clear the workboard holder and the rotary disk. As the head begins the insertion cycle, the cut and clinch anvil is generally raised to a theoretical 0.12 to 0.25 mm (0.005 to 0.010 in) below the board to support it and provide lead clearances.

At the bottom of the head stroke, the cutters are driven upward to cut and clinch the component leads. See Fig. 5.7 for typical minimum side-to-side and end-to-end clearances for the axial-lead cut and clinch to previously inserted components of both axial-leaded packages and DIPs.

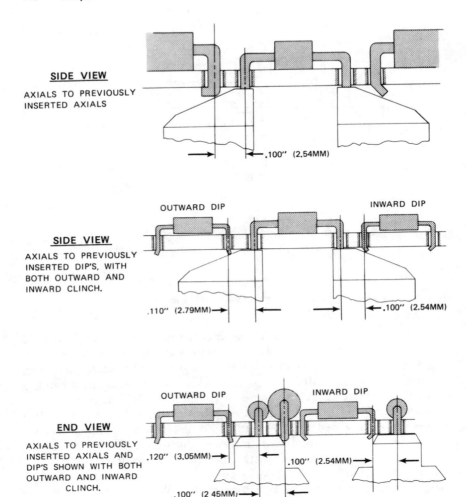

SIDE VIEW

AXIALS TO PREVIOUSLY
INSERTED AXIALS

.100″ (2.54MM)

OUTWARD DIP   INWARD DIP

SIDE VIEW

AXIALS TO PREVIOUSLY
INSERTED DIP'S, WITH
BOTH OUTWARD AND
INWARD CLINCH.

.110″ (2.79MM)   .100″ (2.54MM)

OUTWARD DIP   INWARD DIP

END VIEW

AXIALS TO PREVIOUSLY
INSERTED AXIALS AND
DIP'S SHOWN WITH BOTH
OUTWARD AND INWARD
CLINCH.

.120″ (3.05MM)   .100″ (2.54MM)

.100″ (2 45MM)

Figure 5.7   Minimum axial-lead component clinch clearances.

5.4.8.3 Uninsertable area.   The various methods used to locate boards
for insertion exclude mounting of components, in most cases, in areas
around reference points. This varies greatly between the different
board configurations; however, various techniques can be used to reduce this to an acceptable minimum.

As a general rule, the minimum uninsertable area in the vicinity of
any reference hole is approximately 12.7-mm (0.5-in) radius from the
edge of the workboard holder which supports the locating pin. Edge-
locating and support methods usually require 12.7 mm (0.5 in) along

USING INSERTS AND CHAMFERED EDGES

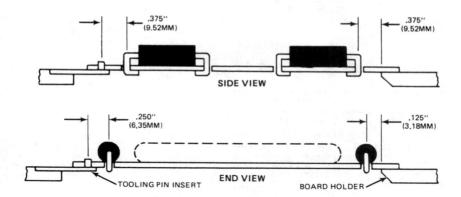

USING BOARD HOLDER—MOUNTED PINS AND NO CHAMFERS

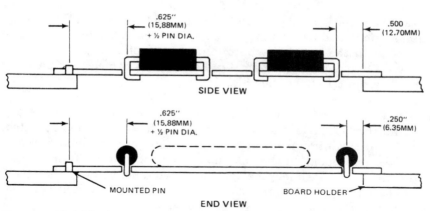

Figure 5.8 Axial-lead component uninsertable areas—clinch to workholder for VCD insertion machines.

the edge guides. With the use of inserts, these requirements can be reduced. See Fig. 5.8. Certain rotary tables have additional uninsertable areas in the four corners. These are usually defined by the data supplied with the specific machine.

5.4.9 Clinch patterns

Lead forming for axial and other similar components is along the centerline of the component, parallel to the component leads, and in-

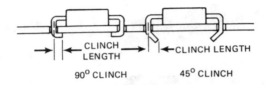

90° CLINCH   45° CLINCH

WIRE DIA.	MIN. CLINCH	MAX. CLINCH
	90° CLINCH	
.020″(0.51MM)	.030″(0.76MM)	.060″(1.52MM)
.030″(0.76MM)	.040″(1.02MM)	.060″(1.52MM)
.040″(1.02MM)	.050″(1.27MM)	.060″(1.52MM)
.050″(1.27MM)	.060″(1.52MM)	.060″(1.52MM)
	45° CLINCH	
.020″(0.51MM)	.020″(0.51MM)	.042″(1.07MM)
.030″(0.76MM)	.028″(0.71MM)	.042″(1.07MM)
.040″(1.02MM)	.035″(0.89MM)	.042″(1.07MM)
.050″(1.27MM)	.042″(1.07MM)	.042″(1.07MM)

Figure 5.9 Typical axial-lead component lead clinch pattern options.

ward toward the body. Although the component is cut from the tape by the insertion head before insertion, a second and more precise cut occurs beneath the board. The minimum length of the lead under the board is a function of lead diameter. Lead length is measured parallel to the board after clinch. See Fig. 5.9.

5.4.9.1 Clinch length. The clinch length is generally measured from the centerline of the lead as it extends through the hole in the printed circuit board. Minimum clinch lengths for smaller wire diameters are somewhat dependent on the hole diameter in the board. To maintain these minimum clinch lengths, hole size should not be more than 0.38 mm (0.015 in) larger than the lead diameter. Clinch length is adjusted by changing anvil span relative to head span.

5.4.9.2 Clinch repeatability. Once the clinch unit is set for a lead length, the repeatability is usually ± 0.13 mm (0.005 in) or less depending on the lead material consistency, with all other insertion parameters remaining constant.

5.4.9.3 Clinch angle adjustment. The clinch may be adjusted to give any desired clinch angle from 30 to 90°. Figure 5.9 shows a 45° and a 90° clinch. When selecting the clinch angle for a given application, hole diameter and lead length should also be considered. Board thickness variations, lead diameter, and material will also have an effect on the clinch angle.

One of the major concerns in selecting an appropriate clinch angle is the solderability of the lead to the printed circuit board. Generally, the 45° clinch is preferred because it is easier to repair than the 90° (tight) clinch; however, there are certain limitations to consider. With single-sided boards which have no holes plated through, the 45° clinch is usually good only if the hole diameter does not exceed the lead wire diameter by, say, 0.38 mm (0.015 in). Generally, this is about the limit for achieving a good solder joint without a void.

On the other hand, there are some users who have used axial components on single-sided boards having no plated-through holes, with hole diameters of 1.5 mm (0.060 in) and a wire diameter of 0.5 mm (0.020 in) with a tight 90° clinch and have felt they have had totally acceptable soldering. With plated-through holes, the above soldering limitations practically vanish regardless of clinch angle and wire-to-hole clearance.

5.4.10   Pattern program considerations

5.4.10.1   Insertion reference point.   The processor-controlled axial-lead insertion machines require that a pattern program be entered into the controller memory. The pattern program contains all the information required to automatically populate a printed circuit board including component location, insertion span, and body diameter.

When generating a pattern program, it is necessary to know the insertion reference point of the component to be inserted. The component insertion reference point of axial-lead components is the intersection of the X (distance between hole centers) and the Y (line running through the center of the component leads) centerlines of the component as inserted into the printed circuit board. In this respect, it is important that initial board layout include positioning holes with respect to a board datum reference. See Fig. 5.10.

5.4.10.2   Insertion center.   The Z axis of processor-controlled axial-lead insertion machines defines the insertion center distance of the component insertion tooling. The insertion center distance (programmed Z span) is the distance between component lead hole centers plus one component wire diameter.

5.4.10.3   Optimum pattern programming.   The pattern program can generally make optimum use of the axial-lead insertion equipment by minimizing X-axis movement. Whenever possible, programming should proceed in a plus or minus Y direction. It is obvious that the distance from "park" to the first component in the "string" and from the last component to "park" again should be minimized. This would

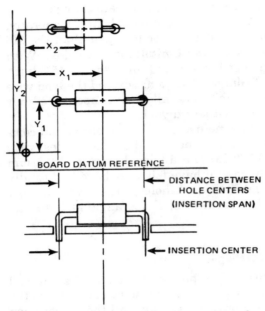

BOARD DATUM REFERENCE

DISTANCE BETWEEN
HOLE CENTERS
(INSERTION SPAN)

INSERTION CENTER

Figure 5.10   Axial-lead component insertion reference point.

be true for both manually loaded machines and pass-through machines.

Where there are a large number of variations in insertion spans (Z moves) in a given board because the Z axis drive is often slower than the X and Y axis drives, this will sometimes lower the yield significantly. Some programmers have found that they can actually increase the overall rate by making two passes over the board; one for the large span (or spans), and the second for the small (or smaller) spans. For example, if the span differences are over 5 mm (0.200 in) and you have a long production run, this is worth a try.

5.5 Radial-Lead Component Insertion

Some of these machines are designed to automatically insert most randomly sequenced radial-lead devices with two leads, and also three-lead transistor devices (Fig. 5.11). Radial-lead sequence inserters are often capable of processing disk and electrolytic capacitors, coils, resistors, thermistors, and similar two-leaded devices packaged in a radial configuration. Other components which can usually be processed and inserted include axial-lead components that have been prepared for processing as radial-lead devices by "hairpin" lead forming,

Figure 5.11   Radial-lead component inserter.   *(Courtesy of Universal Instruments.)*

or with the use of input-forming stations, can be taken directly from standard axial-lead tape. In addition, three-lead transistors that have been prepped for inline lead taping may also be used.

5.5.1   Taping considerations

Component input for some machines is through a component sequencing module. Properly prepared components may be delivered to the dispenser heads from reels and "ammo pack" cartons, which are the preferred methods, or from cassettes.

Radial-lead taped components should meet appropriate lead-to-tape adhesion tests and taped-component-removal pull tests. Tape may be spliced using either an acceptable splicing tape and tool, or by using staples. The use of splicing tape is generally recommended for its adaptability to automatic machine processing. If staples are used, they cannot be placed along the centerline of the feed holes. Placement of staples in the feed hole area will interfere with carrier tape cutting during automatic processing and may damage the dispensing head cutters. For maximum feed reliability, tape feed holes should remain free of splice interference, all feed holes must remain clear of punched material, and the overall tape thickness should not exceed 1.4 mm (0.056 in).

5.5.2 Hole diameter requirements

The minimum hole diameter for radial-lead component insertion should equal the lead diameter plus 0.33 mm (0.013 in). The maximum hole diameter should not exceed the sum of the lead diameter plus 0.38mm (0.015 in). The maximum would be recommended for highest reliability. When printed circuit board density makes it necessary to insert component leads into printed circuit board holes where the two diameters are close, such as 0.36-mm (0.014-in) leads into 0.69-mm (0.027-in) holes or 0.71-mm (0.028-in) leads into 1.04-mm (0.041-in) holes, reliability will be enhanced by considering the jaw-tooling design when laying out the printed circuit board hole drilling centers.

During the insertion process two outer spring clamps are used to grasp the component leads. These clamps secure the outside component leads against two fixed surfaces of the "jaw guide." As a consequence of this clamping method, the actual component lead span will increase or decrease from the minimum insertion span by a factor of one lead diameter plus 4.50 mm (0.177 in) as the lead diameter increases or decreases from optimum.

5.5.3 Board considerations

5.5.3.1 Thickness. The thickness range of board material for radial-lead component insertion is usually identical to that for axial-leaded components. The recommended minimum printed circuit board thickness is 0.81 mm (0.032 in) and the maximum allowable is 2.36 mm (0.093 in) for most machines.

5.5.3.2 Warp and sag. Printed circuit board warp is not easy to define, and its effect varies widely with the type of board handling that is used. For a bare board, warp generally creates a problem during feeding from a stack where automatic board handling is used. With automatic board handling, the problem continues to increase as the weight of inserted components causes a sag because of the typical front and back edge support used in some pass-through-lead machines.

For most radial-lead component insertion machines, the combination of board warp and sag should not exceed 0.1 mm/min (0.004 in/in) with a maximum center deflection in either direction of ($\pm$ 0.062 in) 1.6 mm. It may be necessary to provide some form of board support at or near the centerline to meet these requirements.

For a manually fed machine the same combination of sag and warp still applies, since neither the head tooling nor the clinch is capable of correcting vertical board displacements beyond those dimensions stated. With a workboard holder giving four edge supports, this may completely eliminate any need for an intermediate board support, even on larger board sizes.

5.5.4 Location considerations

5.5.4.1 Above the board. Some radial-lead insertion heads consist of an "insertion jaw" which positions the component and an insert pusher unit which exerts pressure on the top of the component envelope after it is positioned at the programmed hole location. The insert pusher unit is designed to permit overtravel beyond the insertion jaw tip. The insertion jaws may often be rotated + 90°/0°/90° relative to the front of the machine. This motion is controlled by the pattern program. The insertion jaw retracts away from the inserted component while the part is being inserted. Special consideration should be given so that the insertion jaw does not interfere with previously inserted components.

The direction of the insertion jaw as it moves away from the component, as well as the clearance required in relation to particular standard components, is also shown. These clearances when maintained, will prevent component damage during the insertion process.

5.5.4.2 Below the board. The cut and clinch mechanism is generally raised to accept the component leads as they pass through the printed circuit board and to support the bottom of the board to prevent excessive flexing during the insertion process, thereby ensuring a uniform lead length. When the component leads are completely through the printed circuit board, two cutters trim the leads to their finished length and clinch them to the underside of the printed circuit board to secure the component to the board. The unit is then lowered to await the next insertion cycle.

Sometimes two cut and clinch modules are available. The first is often called an "adjustable cut and clinch." This cut and clinch yields a tight component with some sacrifice on longer lead lengths. Any customer who has a further insertion process after the radial-lead devices or expects to experience rough board handling, would experience increased reliability with this unit. The second module is often called "nonadjustable." This cutter head provides a typically shorter lead length that does not vary with lead diameter or material consistency. It is generally recommended for use by those customers requiring specific lead angles and lengths, e.g., military applications.

The compromise involves maintaining close hole tolerances to ensure stable components after insertion. When hole sizes exceed 0.43 mm (0.017 in) over the lead diameter, component stability is often affected, that is, it can move slightly, hence the potential for insertion errors increases since previously inserted components tend to protrude into the insertion head path during the next adjacent cycle.

5.5.4.3 Uninsertable area.

The various methods used to locate boards for insertion exclude mounting of components in areas around reference points, or along the edge guides. Uninsertable area may vary significantly between different board configurations. As a general rule, the minimum uninsertable area in the vicinity of any reference hole is approximately a 15.62-mm (0.625-in) radius.

The areas around the locator arms are further possible uninsertable areas. The determination of particular dimensions is also dependent upon the locator arm assembly and the footprint of the specific machine to be used for component processing.

5.5.5 Cut and clinch patterns

The cut and clinch unit usually rotates in a 0°, 90° counterclockwise relationship with the insertion jaw. The leads of a two-lead device are often clinched in an outward pattern along the axis formed by the centerlines of the leads. The lead clinch angle and trimmed lead length are a function of the clinch selected. While the insertion head may often rotate both ± 90° from the normal position, the cut and clinch unit often rotates only 90° clockwise for either 90° rotation of the head.

The radial-lead cut and clinch unit is mounted below the X-Y table, directly under the insertion-head assembly, and is generally activated in conjunction with the insertion head. The typical cut and clinch unit contains anvils which raise to support the printed circuit board during component insertion. The anvil indexing mechanism usually allows the anvil to be oriented at 0° or 90° in the clockwise direction with reference to the front of the machine. The desired orientation is entered in the pattern program.

5.5.6 Pattern program considerations

Processor-controlled radial-lead insertion machines generally require that a pattern program be entered into the controller memory. The pattern program contains all the information required to populate a printed circuit board, including component locations (X-Y table coordinates of each component insertion), sequencer head number, insertion orientation, and function type. When a printed circuit board is densely populated, the pattern program should be written with the tooling clearances and component orientations taken into consideration.

5.5.6.1 Insertion reference point. When generating a pattern program, it is necessary to know the insertion reference point of the component to be inserted. The insertion reference point for a two- or three-leaded component in a radial configuration is typically the point formed by the intersection of the X (insertion span) and Y (perpendicular line equidistant from the end points of the X insertion span) centerlines of the component.

5.5.6.2 Optimum pattern programming. Generally, in order to make optimum use of the radial-lead insertion equipment and prevent tooling interference with previously inserted components, insertion machines should be programmed with components in the rear of the machine inserted first. Component insertion should then proceed from the back of the board to the front.

The motion of the insertion jaw should be considered when programming. The insertion jaw moves away from the inserted component as the pusher applies pressure. When the jaw is oriented in the 0° position, it will move toward the front of the machine. When oriented 90° in the clockwise direction, the jaw moves to the left, while a 90° counterclockwise orientation results in an insertion jaw movement to the right. Minimum clearance should be considered when a pattern program is written.

5.5.6.3 Special programming considerations. Generally, pattern programs for the radial-lead insertion machines follow the same format as other insertion machines.

5.6 Dual-Inline Package Component Insertion

5.6.1 General

The equipment available to automatically insert DIP components (Fig. 5.12) include insertion systems which are capable of processing one or more of the following devices:

- 7.62-mm (0.300-in) lead span [6 to 20 leads (standard) 16 to 24 leads (optional)]

Figure 5.12 Dual-inline package (DIP) inserter. *(Courtesy of Universal Instruments.)*

- 10.16-mm (0.400-in) lead span (22 to 40 leads)
- 15.24-mm (0.600-in) lead span [22 to 40 leads [standard] or 22 to 42 leads (optional)]
- Two- and four-lead DIP devices with 7.62-mm (0.300-in) lead span

Most insertion systems can be tooled to insert DIP components with lead spans of 7.625, 10.16 or 15.25 mm (0.300, 0.400, or 0.600 in) provided they meet the appropriate DIP dimensions, such as those shown in Fig. 5.13. Components for input to the automatic DIP insertion machine are usually loaded in plastic or metal DIP stocks that are placed into the input magazines of the machine.

The guidelines in this chapter will provide optimum reliability in the production environment. As with all machines, more generous clearances will improve insertion reliability; pushing dimensions to

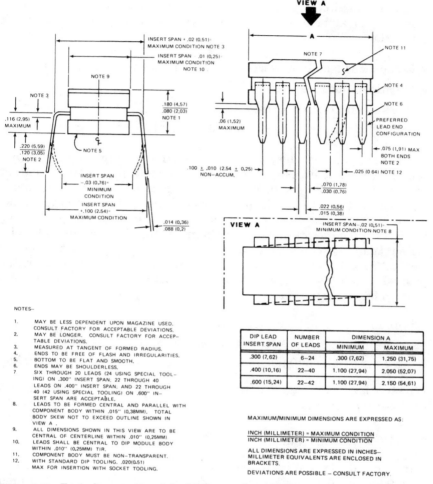

NOTES—

1. MAY BE LESS DEPENDENT UPON MAGAZINE USED. CONSULT FACTORY FOR ACCEPTABLE DEVIATIONS.
2. MAY BE LONGER. CONSULT FACTORY FOR ACCEPTABLE DEVIATIONS.
3. MEASURED AT TANGENT OF FORMED RADIUS.
4. ENDS TO BE FREE OF FLASH AND IRREGULARITIES.
5. BOTTOM TO BE FLAT AND SMOOTH.
6. ENDS MAY BE SHOULDERLESS.
7. SIX THROUGH 20 LEADS (24 USING SPECIAL TOOLING) ON .300" INSERT SPAN, 22 THROUGH 40 LEADS ON .400" INSERT SPAN, AND 22 THROUGH 40 (42 USING SPECIAL TOOLING) ON .600" INSERT SPAN ARE ACCEPTABLE.
8. LEADS TO BE FORMED CENTRAL AND PARALLEL WITH COMPONENT BODY WITHIN .015" (0,38MM). TOTAL BODY SKEW NOT TO EXCEED OUTLINE SHOWN IN VIEW A .
9. ALL DIMENSIONS SHOWN IN THIS VIEW ARE TO BE CENTRAL OF CENTERLINE WITHIN .010" (0,25MM)
10. LEADS SHALL BE CENTRAL TO DIP MODULE BODY WITHIN .010" (0,25MM) TIR.
11. COMPONENT BODY MUST BE NON-TRANSPARENT.
12. WITH STANDARD DIP TOOLING, .020(0,51) MAX FOR INSERTION WITH SOCKET TOOLING.

DIP LEAD INSERT SPAN	NUMBER OF LEADS	DIMENSION A	
		MINIMUM	MAXIMUM
.300 (7,62)	6–24	.300 (7,62)	1.250 (31,75)
.400 (10,16)	22–40	1.100 (27,94)	2.050 (52,07)
.600 (15,24)	22–42	1.100 (27,94)	2.150 (54,61)

MAXIMUM/MINIMUM DIMENSIONS ARE EXPRESSED AS:

INCH (MILLIMETER) = MAXIMUM CONDITION
INCH (MILLIMETER) = MINIMUM CONDITION

ALL DIMENSIONS ARE EXPRESSED IN INCHES—MILLIMETER EQUIVALENTS ARE ENCLOSED IN BRACKETS.

DEVIATIONS ARE POSSIBLE – CONSULT FACTORY.

Figure 5.13 Typical conventional dual-inline package (DIP) component configuration (6 through 42 leads).

the lower limits will reduce insertion reliabilities. Other configurations and dimensions are often possible.

5.6.2 Hole diameter considerations

When determining lead hole diameters, two primary factors must be considered: the first being that the holes are large enough to consistently accept component lead insertion; the second being that the holes are small enough to assure a secure lead clinch. The minimum component lead hole size required for reliable DIP module insertion is a function of

- Component lead cross-sectional area and lead-end configuration
- Machine tolerance and accuracy
- Workboard holder accuracy
- Printed circuit board pattern and tooling reference accuracy

The accumulated tolerances of the insertion tooling, positioning system, and workboard holder normally provide a positioning accuracy of ± 0.12 mm (0.005 in), relative to true position, for a machine that is equipped with a standard, rotary positioning system. In order to determine the minimum acceptable lead hole diameter, this positioning accuracy often dictates a starting diameter that is 0.25-mm (0.010-in) larger than the effective lead diameter of the component to be inserted. To this must be added the manufacturing tolerance of the printed circuit board hole pattern (hole location tolerance). This is expressed by the equation

Minimum hole diameter = effective lead diameter

+ hole location tolerance

+ 0.010 inch (0.008 inch for machines not using a rotary table)

Effective lead diameter is a function of the tooling lead slot (0.056-mm [0.022-in]) and component lead thickness, where

$$\text{Effective lead diameter} = \sqrt{(0.022 \text{ in})^2 + (\text{lead thickness})^2}$$

In general, the minimum component hole diameter should be less than 0.037 in (0.94 mm), except as described in Sec. 5.6.3 for pointed or tapered leads, nor should the maximum hole diameter exceed 1.03 mm (0.041 in).

If a rotary table feature is not being used, the minimum component hole diameter can usually be safely reduced to 0.94 mm (0.037 in) as the positioning systems accuracy now becomes ± 0.1 mm (0.004-in). Whenever inserting into printed circuit board holes smaller than those specified in this paragraph is being considered, the equipment manufacturer should be consulted.

5.6.3 Component lead considerations

Insertion reliability for DIP modules is also influenced by the condition of the component leads as well as by the lead tip configuration. Sharp bends in the leads will prevent their being inserted. Leads terminated in a point will enhance insertion reliability and can effectively reduce the required hole diameter by 0.008 mm (0.003 in) or 0.15 mm (0.006 in) depending upon the degree of taper, length of taper, lead material, and printed circuit board hole plating.

5.6.4 Component mix

Ideally, the best production rates are obtainable when using components which have identical physical characteristics and tolerances. Because this is not always possible, it must be noted that the greater the range and the greater the variety of deviations there are between the component types being processed, the greater the probability that stated insertion rates and reliability will be adversely affected. Where either or both of these conditions are anticipated, the factory should be consulted to assure that the best ratio between component mix and productivity/reliability is obtained.

5.6.5 Location considerations

5.6.5.1 Above the board. In general, where axial-lead components and DIP modules are intermixed on the same board, the DIP modules should usually be inserted first and the spacing rules for axial leads should be applied provided that the axial-lead component height is not greater than 6.35 mm (0.250 in) at the point of clearance. For end-to-end and side-to-side spacing requirements for the DIP modules, refer to Fig. 5.14.

5.6.5.2 Below the board. The clinch-down position is below the workboard holder providing clearance for positioning movement. As the insertion head begins the insertion cycle, the cut and clinch anvil is raised to support the printed circuit board during the component lead insertion. At the bottom of the head stroke and after the insertion plunger is extended, the cutters are driven in or out, depending upon pattern program selection, to cut and clinch the component leads. The workboard holder edges should not be positioned so as to obstruct the operation of the cut and clinch assembly. This restriction is taken into consideration in the dimensions presented in Fig. 5.15.

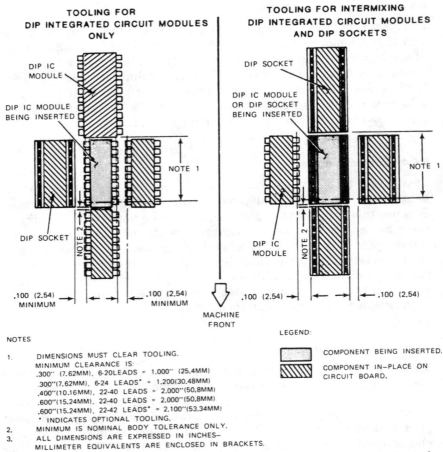

Figure 5.14 Typical dual-inline package (DIP) component-to-component insertion tooling clearances.

5.6.5.3 Uninsertable area.

The various methods used to locate printed circuit boards on the workboard holder generally do not permit components to be located in areas around board reference points. Uninsertable area varies greatly between the different workboard holder configurations, however. Various design techniques can be used to reduce this uninsertable area to an acceptable minimum.

As a general rule, the minimum uninsertable area in the vicinity of any reference hole location is approximately a 25-mm (1-in) radius. Edge-locating and support methods usually require a minimum of 25-mm (1-in) clearance along the guide edges. Use of inserts can reduce these requirements, see Fig. 5.15.

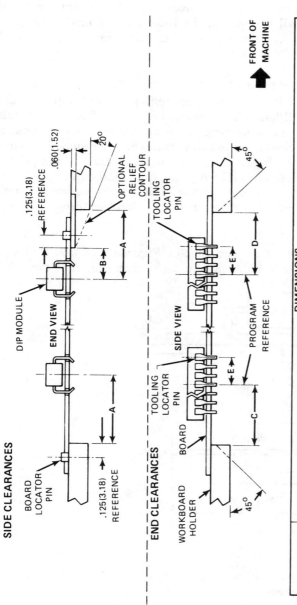

SIDE CLEARANCES

BOARD LOCATOR PIN

DIP MODULE

.125(3.18) REFERENCE

.060(1.52) REFERENCE

.125(3.18) REFERENCE

OPTIONAL RELIEF CONTOUR

20°

END VIEW

END CLEARANCES

TOOLING LOCATOR PIN

BOARD

WORKBOARD HOLDER

TOOLING LOCATOR PIN

PROGRAM REFERENCE

SIDE VIEW

45°

45°

FRONT OF MACHINE

DIP LEAD INSERT SPAN	A	B	DIMENSIONS				E
			WITHOUT 45° RELIEF		WITH 45° RELIEF		
			C	D	C	D	
.300(7.62)	1.550(39.37)	.940(23.88)	.980(24.89)	.680(17.27)	.744(18.9)	.444(11.28)	.300(7.62)
.400(10.16)	1.600(40.64)	.990(25.15)	1.480(37.59)	1.180(29.97)	1.244(31.6)	.944(23.98)	.800(20.32)
.600(15.24)	1.700(43,18)	1.090(27.69)	1.480(37.59)	1.180(29.97)	1.244(31.6)	.944(23.98)	.800(20.32)

ALL DIMENSION ARE EXPRESSED IN INCHES, METRIC EQUIVALENTS ARE BRACKETED.

Figure 5.15 Typical dual-inline package (DIP) uninsertable areas.

5.6.6 Clinch patterns

5.6.6.1 Multiple-center cut and clinch (inward). Multiple-center cut and clinch units will cut all leads from the inserted DIP module. Figure 5.16 shows the typical specifications for this unit when it is configured to clinch in the inward direction.

5.6.6.2 Multiple-center cut and clinch (outward). When used in the outward clinch configuration, the multiple-center cut and clinch units will trim and clinch all leads. The outward clinch configuration tends to provide better component seating by taking advantage of the natural outward spring of the leads. It also permits easier removal of parts from the printed circuit board in the event replacement is required. Specifications for the outward clinch pattern are shown in Fig. 5.17.

Selective lead clinch patterns are available for the outward clinch configuration. However, there are insertion density trade-offs that must be considered.

5.6.6.3 Two- and four-lead cut and clinch. Figures 5.16 and 5.17 can also apply to the two- and four-lead cut and clinch unit. The two- and

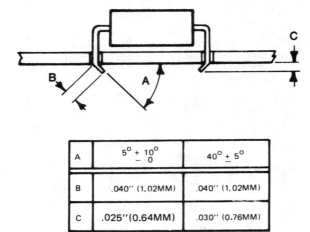

A	$5^{\circ} \begin{array}{l} + 10^{\circ} \\ - 0 \end{array}$	$40^{\circ} \pm 5^{\circ}$
B	.040" (1.02MM)	.040" (1.02MM)
C	.025"(0.64MM)	.030" (0.76MM)

NOTES:
1. TOOLING CHANGE IS REQUIRED TO CHANGE LEAD CLINCH ANGLE.
2. 40° CLINCH ANGLE IS NOT RECOMMENDED FOR SOCKET INSERTION.
3. ANGLE MAY VARY DEPENDING ON LEAD HARDNESS AND HOLE DIAMETER.

Figure 5.16 Typical dual-inline package (DIP) component inward lead clinch.

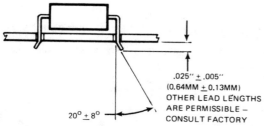

.025" ± .005"
(0.64MM ± 0.13MM)
OTHER LEAD LENGTHS
ARE PERMISSIBLE –
20° ± 8°   CONSULT FACTORY

Figure 5.17   Typical dual-inline package (DIP) component outward lead clinch.

four-lead cut and clinch unit clinch configuration (inward or outward) is also usually programmable.

5.6.7   Programming considerations

5.6.7.1   Insertion reference.   When the DIP module is positioned against mechanical stops in the "spreader pusher" assembly, the two-DIP module leads nearest the front of the machine are used as locators to ensure that the DIP module is in the proper position for pickup and insertion. The programming reference point is usually determined relative to these same two leads. The component reference point, or programming reference point, is the imaginary point on the DIP module body relative to which the X and Y coordinates are defined. The programming reference point depends upon the component lead insertion span.

For 7.62-mm (0.300-in) DIP modules, the programming reference point is an imaginary point 7.62 mm (0.300 in) from the locating leads and centerline of insertion. For 10.16-mm (0.400-in) and 10.16 and 15.24 mm (0.400 and 0.600 in) DIP modules, the programming reference point is located 20.3 mm (0.800 in) from the locating leads and on the centerline of insertion.

5.6.7.2   Magazine assignments.   Reference should be made to the specific machine general specification and manual to determine magazine pattern programming information.

5.6.7.3   Optimum pattern programming.   The pattern program can make optimum use of the DIP module insertion equipment by minimizing Y axis movement. Whenever possible, programming should proceed in a plus or minus X axis direction. In order to prevent tooling interference with previously inserted components, the pattern program should call for the insertion of front-row components first, and insertion should proceed toward the back row. This is only necessary if close front-to-back spacing is used.

References

1. "Design Guidelines for Leaded Component Insertion," Universal Instruments Corporation, Binghamton, N.Y.
2. "Reel Packaging of Components with Axial Leads," EIA RS-296, Electronic Industries Association, Washington, D.C.

Flexible Printed Circuit Design

6.1 Introduction[1]

Before getting into the specifics of flexible printed circuit design it should first be determined if a suitable application exists. There are several important considerations behind the design of most flexible printed circuits. The most significant of these are

- *Cost:* The total installed cost of the functioning interconnection should be considered, not simply the price of circuitry as compared with wire.

- *Function:* An evaluation of the circuit's purpose and whether that purpose could be served as well by other wiring techniques should be made.

- *Wiring:* A capability should be examined for mass circuit termination to save production time and eliminate wiring errors.

6.2 General Considerations[2]

There are many factors that affect the design of a flexible printed circuit and major cost advantages result from design. The selection of standard materials and thicknesses and all noncritical tolerances designed loose enough to allow automated production are two important considerations.

There are two general sets of concerns in designing flexible printed circuits: electrical and mechanical. Several of these issues also pertain to rigid printed circuit boards as already discussed in this book. Thus, this chapter will concentrate on those issues that are more closely allied with the design of flexible printed circuits.

6.2.1 Design specifications[3,4]

In general, depending on the end-product application, most flexible printed circuits are designed in accordance with the requirements of either IPC-D-249, "Design Standard for Flexible 1 and 2 Sided Printed Boards," or MIL-STD-2118, "Design Requirements for Flexible and Rigid-Flex Printed Wiring for Electronic Equipment."

6.2.2 Selection criteria

The use of flexible printed circuits is not always the answer for an application. Flexible printed circuits should, however, be considered if it is necessary to:

- *Reduce package size and weight:* Flexible printed circuits weigh less, take up less space, and can be formed to fit in very small spaces.

- *Reduce assembly costs and wiring errors:* Flexible printed circuits predictably and reliably eliminate human discrete wiring errors.

- *Improve cost effectiveness:* The use of flexible printed circuits is often less expensive than discrete wiring when more than 20 point-to-point connections are being made. Below this, depending on complexity and quantity, point-to-point discrete wiring is generally less expensive.

- *Enhanced flexural endurance:* The flat conductor construction of flexible printed circuits enhances the flexural endurance ("flex-life") of continuously or intermittently flexed or bent interconnection wiring systems.

- *Tightly controlled electrical performance:* Since the raw materials are very consistent as to dielectric thickness this simplifies the production of controlled-impedance transmission lines and provides repeatable performance.

- *Improved reliability:* When used to reduce the number of interconnections in a system, a nearly-automatic result is the increased reliability of the end-product equipment.

- *Compliant mounting:* Flexible laminates provide a compliant substrate that mitigates the issue of mismatched coefficients of thermal expansion (CTE) that is often a concern when designing surface-mount assemblies.

Flexible printed circuits are typically more expensive than their rigid counterparts. However, packaging systems that use flexible printed circuits generally have a lower "out-the-door" cost and better field reliability than hard-wired alternatives. Thus, a holistic approach should be taken when designing flexible printed circuits because the interrelationship of materials, manufacturing methods, reliability requirements, and ultimate product cost are intrinsically linked to the design.

6.3   Types of Flexible Printed Circuits[5]

There are several types of flexible printed circuits, as listed in Table 6.1. The simplest do not require electrical interconnections between layers; the more complex do. Plated-through holes are most often used to provide interconnections between layers. For purposes of design the plated-through hole interconnection approach is the same as is used in conventional rigid multilayer circuits. The only difference is that in flexible circuitry the through-hole should be located in an area that will receive limited flexing. Accordingly, it is often desirable specifically to rigidize the areas of the circuits that contain through-holes.

6.3.1   Single-sided flexible printed circuits[6]

A flexible printed circuit that is considered to be "single-sided" can in itself take many forms beyond what is normally considered in rigid printing wiring. For purpose of definition, all flexible printed circuits that contain a single conductor layer will be considered to be of this type. In general, single-sided flexible printed circuits should be in accordance with the requirements of IPC-FC-240, "Specification for Single-Sided Flexible Printed Wiring."

Single-sided flexible printed circuits can further be divided into subclasses. These subclasses are as follows:

- Single-access uncovered
- Single-access covered
- Double-access uncovered
- Double-access covered

The flexible printed wiring can be formed at assembly to conform to a three-dimensional shape, eliminating the need for wire cable to make the terminations, or the printed wiring itself can be formed during its manufacture for conformance or may be made to be self-extending. Termination of the printed wiring can be accomplished by conven-

TABLE 6.1   Types of Flexible Printed Circuits

Type	Advantages	Limitations	Relative Cost
Single-Sided Access Area— No Dielectric, Base Layer, Base Layer, Conductors, Conductors, Cover Layer	Least Expensive	Conductor Crossovers Not Possible	1
Double-Sided Top, Base, Conductors, Plated Through Hole, Bottom	Higher Density — Reduced Number of Solder Joints	Increased Cost — Reduced Flexibility	1.5-2
Double Access Base, Hole in Base to Expose Pad from Other Side	Minimum Number of Parts — Component Mounting on Both Sides	More Expensive than Single Access	1.2-1.7
Multilayer Plated Through Hole, Access Hole, Conductors Between Insulation Layers	High Conductor and Component Density — Integral Shielding — Controlled Impedance	High Cost — Reduced Reliability — Cannot Be Repaired — Not Flexible	3 to 20*
Rigid-Flex Several Layers of FPC Not Banded, Rigid, Multilayer Wiring	Combines Multilayer Features with Selective Flexibility.	(Same as Multilayer)	3 to 20*
Rigidized FPC FPC, Stiffeners (No Wiring)	Strain Relief and Component Support	Solder Access. Restricted to One Side	1.2-4

*Depends on number of layers

SOURCE: Sheldahl Corporation.

tional printed wire methods or many systems that were exclusive to wire such as spade lug screw connections. Designs are restricted only to the imagination of the designer. Such an example as forming a portion of the flexible wiring in a U shape so as to function as the contacts of a connector has been used in such stringent application as missile work.

6.3.1.1 Single-access uncovered. Single-access covered flexible printed circuits are the most inexpensive type. They consist of a dielectric layer which supports the conductor. This type of flexible printed circuit finds most application where the wiring will not be exposed to mechanical abuse or environmental contamination. The components are normally all mounted from the dielectric side with all electrical connections to the conductor side. The connections can be soldered, welded, or pressure type. This type of wiring has found great acceptance in such items as telephones, relays, toys, small motors, and many other such small enclosed assemblies.

6.3.1.2 Single-access covered. Single-access covered flexible printed circuits consist of a three-layer structure—two dielectric layers with a conductor layer sandwiched between. The dielectric is selectively removed from those areas of the conductor to which subsequent electrical connection is to be made.

The cover lay added to the basic wiring gives the advantage of a moisture and insulation barrier for conductor-to-conductor spacing, and insulation from associated mounting chassis or hardware. The cover lay having oriented holes does have a significant effect on cost.

6.3.1.3 Double-access uncovered. Double-access flexible printed circuits have the unique feature of electrical access to the single layer conductor from both sides. The structure consists of two layers, dielectric and conductor. The dielectric layer is fabricated with holes, while the conductor pattern is suspended across the holes allowing contact from either side of flexible printed wiring. This type of flexible printed circuit has been most frequently used in high-production-quantity applications where inexpensive electrical connection is required with a minimum amount of components, but mounting is necessary from both sides. Dedicated production tooling is usually required for this type of circuit.

6.3.1.4 Double-access covered. Double-access flexible printing circuits are made up of three layers—two dielectric layers and a single conductor layer. The conductor layer is sandwiched between the two insulating layers which have selective access holes to the conductor. The access holes may be at opposing points, leaving the bare conductor exposed on both sides, or the access areas may be selectively located on either side.

The application for this type of printed circuit is the most versatile single-sided type. The cover lay not only can serve the purpose of insulation from associated hardware but can also serve to insulate from itself. Many applications of point-to-point wiring as well as component mounting have been developed with this configuration.

6.3.2 Double-sided flexible printed circuits[7]

As the name implies, double-sided flexible printed circuits have conductors on both sides of a base dielectric material. Conductors on opposite sides may be connected by plated-through holes or other means. In this and other respects they are very similar in construction to rigid double-sided printed circuit boards.

Double-sided (and multilayer) flexible printed circuits should always be given serious consideration when an application's interconnections become very dense. However, in designing flexible printed circuits with more than one conductor layer, the required degree of flexibility must be kept in mind. Plating, conductor routing, and material thickness can also affect printed circuit flexibility.

Most double-sided flexible printed circuits have plated-through holes. The most economical way to manufacture plated-through holes in flexible printed circuits is to use the "pattern plating" process. However, if end-product flexing or sharp bending is involved, the use of pattern plating might not be recommended. It might be better to mask the conductor pattern and plate only the lands. While this is more costly, in some applications it may be necessary to plate the flexible printed circuit using this technique.

6.3.3 Multilayer flexible printed circuits

The multilayer type of flexible printed circuit is an end product consisting of three or more conductive layers on flexible dielectric bases bonded to form a monolithic mass. An insulating layer may be applied over the outside conductor paths. The multilayer structure in its simplest form is a three-conductor layer structure with the center layer being discreet wiring paths and both outer conductor layers a solid copper conductor. In this type of multilayer the wiring paths are electrically equivalent to coaxial or shielded wire configurations.

The multilayer printed wiring can be by definition an infinite number of conductor layers and is limited only by economics. It should be recognized that for each layer of multilayer structure there is additional dedicated manufacturing tooling and a cumulative cost resulting from manufacturing yields. The most frequently used structure of multilayer construction is a four-layer structure utilizing the adjacent inner layers for power and ground distribution and the outer layers for discreet interconnection between layers.

Many designers have found that in utilizing this concept, a significant savings in cost and manufacturing can be achieved. This is accomplished by a standard configuration of the ground plane and power plane layers which can be produced on a large-quantity basis, and the total structure laminated to a point of solid sheet conductors for the single plane layers. The fabrication of the signal plane paths and interconnection between layers is then only a process equivalent to fabricating conventional plated-through hole double-sided printed circuit board.

This design concept has found great acceptance in the intraunit wiring applications. It provides a constant wiring path of relative short length, with each unit having the same electrical transmission characteristics.

6.3.4 Rigid-flex printed circuits[8,9]

Rigid-flex printed circuits are similar to conventional multilayer printed circuit types except that the bonding and connections between layers are confined to specific areas of the interconnection wiring and component mounting plane. Between the rigid laminated areas, each conductor layer is usually bonded to a single, thin base laminate area that remain flexible even though there may be more than one conductor layers.

In general, rigid-flex printed circuits should be in accordance with the requirements of ANSI/IPC-RF-245, "Performance Specification for Rigid-Flex Printed Wiring Boards." Thus, a rigid-flex printed circuit is in general a combination of flexible and rigid printed circuit elements that are combined through lamination and plating into a single inseparable structure. One of the most important benefits of this type of construction is a reduction in the number of discrete solder joints or connector contacts in the conductive path. Other benefits are savings in space and weight, and lower packaging assembly costs, as termination points are preestablished and eliminate the chances or interconnection wiring errors. A typical rigid-flex circuit and its application are shown in Fig. 6.1.

6.4 Materials[5]

The properties of materials selected for use in circuits are equally as critical as design factors. Thus, special attention should be given to the selection of the appropriate conductive (copper) foil, dielectric materials, and adhesives.

6.4.1 Conductive foil

Copper is by far the most common conductive foil material for flexible printed circuits. Two general types of copper are used, namely, electrolytically deposited (ED) and rolled-annealed (RA). The differences between the two types of copper are

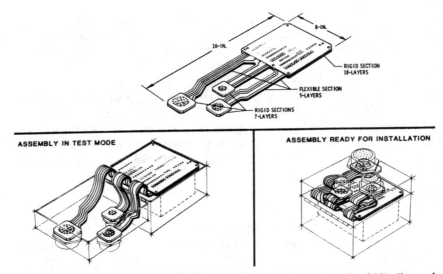

Figure 6.1 Rigid-flex printed circuit application. *(Courtesy of Lockheed Missiles and Space Company.)*

- ED copper has a vertical grain structure, whereas, RA copper has a horizontal grain structure that is better for continued operational flexing.
- RA copper is available in 24-in (61-cm) widths or less.
- ED copper is plated-up and, thus, it is more economical to use on a per unit price basis.
- ED copper has a rougher base surface that enhances adhesive bonding to the base dielectric material.

Copper foil is commercially available in several sizes as listed in Table 6.2. The thickness is generally given as it relates to the copper weight (in ounces) per square foot of foil. Thus, the foil can be as small as ½ oz and as thick as 6 oz. However, the 1- and 2-oz thicknesses are the most commonly used in flexible printed circuits.

6.4.2   Dielectric materials[10,11]

Several different dielectric materials are available for use as the base and cover lay materials in a flexible printed circuit, as given in Table 6.3. However, the two most commonly used dielectric materials are polyester and polyimide films. In general, the base dielectric material should be in accordance with the requirements of ANSI/IPC-FC-231, "Flexible Bare Dielectrics for use in Flexible Printed Wiring." The

TABLE 6.2 Copper Foil Parameters

Foil weight, oz/ft^2	Flat conductor dimensions, in		Cross Section		AWG wire size (based on equiv. current rating)	Resistance, (mΩ/ft at 20°C)	Current for 10°C rise*, A	Current for 30°C rise, A
	Thickness	Width	sq mils	mils				
2	0.0027	0.030	81	102	28	100	2.0	3.4
	0.0027	0.045	122	154	27	67	2.6	3.8
	0.0027	0.060	162	204	25	50	3.1	5.1
	0.0027	0.075	202	254	24	40	3.7	5.8
	0.0027	0.090	243	306	23	34	4.2	6.5
	0.0027	0.125	338	425	22	24	5.4	8.2
	0.0027	0.155	418	527	21	19.5	6.4	9.2
	0.0027	0.185	500	630	20	16.2	7.2	10.7
	0.0027	0.250	675	850	18	12	8.5	13.5
3	0.004	0.030	120	141	27	67	2.3	4.0
	0.004	0.045	180	227	25	45	3.0	5.2
	0.004	0.060	240	302	24	34	3.7	6.0
	0.004	0.075	300	378	23	27	4.4	7.0
	0.004	0.090	360	454	22	22.5	5.0	7.8
	0.004	0.125	500	630	20	16.2	6.2	10.0
	0.004	0.155	620	780	19	13	7.3	11.8
	0.004	0.185	740	920	18	11	8.2	13.5
	0.004	0.250	1000	1260	17	8	10.0	17.0
4	0.0055	0.045	248	312	24	33	3.4	6.0
	0.0055	0.060	330	415	23	25	4.1	7.2
	0.0055	0.075	412	520	22	20	5.0	8.2
	0.0055	0.090	495	624	21	16.5	5.6	9.5
	0.0055	0.125	687	865	20	12	7.1	12.2
	0.0055	0.155	852	1075	19	9.5	8.1	14.8
	0.0055	0.185	1020	1285	17	8	9.2	17
	0.0055	0.250	1375	1370	16	6	11.1	21

TABLE 6.2 Copper Foil Parameters (Continued)

Foil weight, oz/ft²	Flat conductor dimensions, in		Cross Section		AWG wire size (based on equiv. current rating)	Resistance, mΩ/ft at 20°C	Current for 10°C rise*, A	Current for 30°C rise, A
	Thickness	Width	sq mils	mils				
6	0.008	0.045	360	454	22	23	3.7	7.8
	0.008	0.060	480	605	21	17	4.7	9.8
	0.008	0.075	600	755	20	13.5	5.5	11.5
	0.008	0.090	720	905	19	11.2	6.2	13.2
	0.008	0.125	1000	1260	17	8	7.7	17
	0.008	0.155	1240	1560	16	6.5	8.7	20
	0.008	0.185	1480	1860	14	5.5	10.0	23
	0.008	0.250	2000	2520	13	4.1	12.0	26

*Rise from 20°C ambient temperature
SOURCE: Sheldahl Corporation.

TABLE 6.3 Dielectric Material Properties

	TFE glass cloth	FEP glass cloth	Polyimide film	Epoxy dacron	Polyester film	Polyethylene
Specific gravity	2.2	2.2	1.42	1.38	1.395	0.93
Flammable	No	No	Self-exting.	Yes	Yes	Yes
Appearance	Tan	Tan	Amber	Translucent	Clear	Clear
Bondability with adhesives	Good*	Good*	Good	Good*	Good	Poor
Bondability to itself	Poor	Good	Poor	Good	Poor	Good
Chemical resistance	Excellent	Excellent	Excellent	Good	Excellent	Good
Sunlight resistance	Excellent	Excellent	Excellent	Fair	Fair	Low
Water absorption, %	0.10/0.68†	0.18/0.30†	3 per 24 hr	1.5	0.8 per 24 hr	0.01 per 24 hr
Volume resistivity, Ω cm	10^{16}	10^{16}	10^{16}	1×10^{16}	10^{16}	10^{16}
Dielectric constant, 10^2–10^8 Hz	2.5/5†	2.5/5†	3.5		2.8–3.7	2.2
Dissipation factor, 10^2–10^8 Hz	7×10^{-4} / 10^{-3}†	10^{-4} / 10^{-3}†	3×10^{-3} / 14×10^{-3}	70×10^{-3}	2×10^{-3} / 16×10^{-3}	6×10^{-4}
Service temperature						
Min. (°C)	–70	–70	2–00	–20	–60	–20
Max. (°C)	250	200	300	100	105	60
Tensile strength, lb/in² @ 25°C	20,000†	20,000†	25,000	300	20,000	2000
N/m² × 10^{-6} @ 25°C	1.378	1.378	1.697	0.8825	1.378	0.1378
Modulus of elasticity						
lb/in²	4.0†	4.0†	510,000	12,000	550,000	50,000
N/m² × 10^{-6}	0.0003	0.0003	34.623		37.895	3.445
Thermal expansion	Low†	Low†	(–14–38°C)	(–30°C–30°C)	(21–50°C)	(–30–30°C)
in/in/°F × 10^{-6}			11	11	15	100
cm/cm/°C × 10^{-6}			20	20	27	180
Dielectric strength, V/mil	650–1600†	650–1600	7000	900	700	585
Sample size, in	0.003	0.003	0.002		0.001	0.125

*Must be treated.
†Depends on % glass cloth.
SOURCE: Hughes Aircraft Company.

foil-clad laminate should be in accordance with the requirements of ANSI/IPC-FC-241, "Flexible Metal-Clad Dielectrics for Use in Flexible Printed Wiring." The basic factors to consider in choosing a dielectric material are cost, dimensional stability, and their thermal properties, which are listed in Table 6.4.

6.4.3 Adhesives[5]

A flexible printed circuit has distinct adhesive layers that bond it together. The important properties to consider when selecting an adhesive (Table 6.5), are

- Adhesion
- Flexibility
- Chemical resistance
- Moisture absorption
- Thermal resistance
- Electrical properties
- Relative cost

In general, the adhesives should be chosen in accordance with the requirements of ANSI/IPC-FC-233,[12] "Flexible Adhesive Bonding Films." The combination of the adhesive onto a base dielectric should be in accordance with the requirements of ANSI/IPC-FC-232,[13] "Adhesive Coated Dielectric Films for Use as Cover Sheets for Flexible Printed Wiring."

6.5 Design Practice

The following are, in short form, major items of concern, with definitions and reasons why each item should be considered at the earliest stages of the design layout.

6.5.1 Mock-up

A dimensionally stable "paperdoll" mock-up should be used to define, in two dimensions, the physical layout of the flexible printed circuit. This can be subsequently used to expose, early in the design process, assembly difficulties and physical interferences within the package.

6.5.2 Copper biasing

It is a good practice to favor having most of the flexible printed circuit area in copper, even if it is not electrically functional. This bias to-

TABLE 6.4 Dielectric Material Selection Parameters

Cable characteristics	Cable material									
	Polyimide adhesive	Polyimide FEP adhesive	Teflon FEP	Teflon FEP-C	Polyester adhesive	Polyimide epoxy adhesive	Teflon TFE	Acrylic adhesive	Nomex acrylic	Dacron epoxy
Construction Bond										
Fusion Bond			X	X						
Adhesive Bond	X	X		X	X	X	X	X		
Environment										
Service temp., °C	1*	4	5	8	9	6	2	3	6	5
Radiation absorption	1	4	10	10	3	2	5	2	2	4
Physical Factors										
Dimensional stability	4	3	8	5	2	1	7	1	5	4
Flexibility	9	7	1	1	4	8	3	3	1	7
Circuit thickness 6 mm (0.025 in)	4	6	8	2	1	3	8	2	4	8
Stripping ease	10	6	1	4	3	8	2	4	5	10
Land baring ease	10	6	1	2	2	9	NR	4	4	6
Solderability	1	3	6	7	9	2	1	1	4	2
Tear resistance	10	8	1	2	3	9	NR	8	9	5
Welding through ease	NR†	NR	3	3	2	NR	4	NR	NR	NR
Potting bondability	3	3	6	3	2	3	6	2	4	3
Cost Factors										
Raw material	9	10	5	5	3	8	2	8	3	5
Continuous cable (without connectors)	NR	9	7	6	2	10	3	4	4	2
Flexible printed wiring without connectors	10	6	9	3	1	5	NR	5	4	5

*Scale of 1 to 10 ranges from best to least acceptable.
†NR = no results.
SOURCE: Sheldahl Corporation.

TABLE 6.5 Adhesive Selection Parameters

Type	Temperature resistance	Chemical resistance	Electrical properties	Adhesion	Flexibility	Cost	Moisture absorption
Polyester	Fair	Good	Excellent	Excellent	Excellent	Low	Fair
Acrylic	Very good	Good	Good	Excellent	Good	Moderate	Poor
Modified epoxy	Good	Fair	Good/excellent	Excellent	Fair	High	Good
Polyimide	Excellent	Very good	Good	Very good*	Fair	Very high	Poor†
Fluorocarbon	Very good	Excellent	Good	Very good*	Excellent	Moderate	Excellent
Butyral phenolic	Good	Good	Good	Good	Good	Moderate	Fair

*Very difficult to process.
†Evolves water while curing.
SOURCE: Sheldahl Corporation.

ward the use of extra copper tends to counteract the shrinkage that normally occurs as the material relaxes where the copper has been removed (etched).

6.5.3 Conductor routing through bends

It is a good design practice to make sure that conductors in bend areas are routed perpendicular to the bend line. This will enhance flexural life by minimizing conductor bending stresses.

6.5.4 I-beaming

The routing of conductors directly opposite one another on different sides of the dielectric material is called "I-beaming" and should be avoided. The use of I-beaming is particularly detrimental if it occurs in bend areas, because it generally causes premature fatigue failure of the copper. The solution is to either stagger the conductors from one side to the other or move all of the conductors to a single layer in the bend area.

6.5.5 Neutral axis

The physical center of any material, as it applies to bending, is referred to as the "neutral axis." The neutral axis, in theory, will see no stress as a part is flexed. Thus, it is a good design objective to place copper at the exact center of the flexible printed circuit construction in bend areas whenever possible. The best design, especially for dynamic flexing applications, would have only a single layer of copper in the bend areas.

The accepted rules of thumb for bend radii are that they should be from 3 to 6 times the thickness of the flexible printed circuit in the bend area for single-sided types; from 10 to 15 times (or more depending on layer count) for multilayer types; and, from 20 to 40 times, or more, for dynamic bending applications, depending on the number of flexing cycles-to-failure required for the end-product application.

6.5.6 Anchoring spurs

Anchoring spurs, also referred to as "tie-down tabs" or "rabbit ears" are metal projections around a land. These features are fully captured by the cover lay (cover coat) that is normally applied to the top of the flexible printed circuit. The net effect is to mechanically restrain the lands to prevent them from lifting from the base laminate, as shown in Fig. 6.2. This is especially helpful to provide resistance to lifting during initial assembly and field repair soldering operations. The technique is of primary benefit in a single-sided design. However, a word of caution is necessary on the use of anchoring spurs. Anchoring

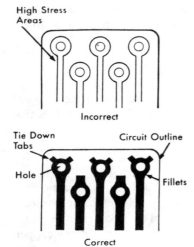

Figure 6.2 Anchoring spurs and fillets. *(From IPC-D-330.)*

spurs used on gigahertz-frequency designs can act as miniature antennae that transmit unwanted electrical noise.

6.5.7 Filleting

The practice of gradually necking-up ("filleting") a conductor when it joins a land moves a potentially high stress point away from the hole edge to a location where it is better protected (see Fig. 6.2). This net effect can also be accomplished by the use of elongated, oval, square, or oversized lands that allow for 360° land capture by the cover lay where space allows.

6.5.8 Access holes

Access holes are openings in the cover lay film that allow access to the land beneath for purposes of either soldering components or allowing for electrical contact. The size of the access hole is generally dictated by the type of circuit being designed and the size of the land. For example, a single-sided flexible printed circuit with fillets and anchoring spurs would require an access hole that is from 0.010 to 0.015 in (0.25 to 0.38 mm) smaller than the land in order to assure that the land will not lift during soldering. Of course, the amount of exposed solderable area is significantly reduced.

Generally, a double-sided flexible printed circuit requires only filleting. Thus, the access hole need only be equal to or greater than the land size. This is due to the "riveting" effect of the plated-through hole.

Most conductive patterns are insulated on both sides. Access holes must, therefore, be designed through one or both sides of the insulation. Access holes should overlap the lands as much as possible, providing additional holddown for the lands. In this way, they strengthen the conductors; not just for the original solder connection but for as many as 10 repair cycles. The least expensive circuit has all access windows on the same side, although product designs do not always allow for single-sided access.

Access to the land can be made by chemical milling in some materials, such as polyamids or polyimides. Other insulations and Teflon films must be end-milled. Accuracy of hole size and location is usually best with materials that have been drilled by numerically controlled equipment.

6.5.9 Tear restraints

Tear restraints are design features incorporated to minimize the risk of tearing the flexible printed circuit. Examples of some methods used successfully as tear restraints include:

- Large internal radii at corners (not square corners)
- Superfluous copper in corners
- Glass cloth or reinforced laminate in corners
- Relief slots
- Holes at the ends of slits
- Tangentially radiused corners
- Cover lay laminated in corners

6.5.10 Strain relief

Mechanical strain-relief reinforcements should be used to prevent or minimize stress concentration on installed flexible printed circuits. Techniques to achieve this include the use of

- Stiffeners bonded to the stress area
- Encapsulation
- Fillet formation at the interface
- Mechanically attached strain-relief bars

6.5.11 Stiffeners

Stiffeners are generally rigid material sections of glass-reinforced plastic laminate, although unfilled plastic, metal, or additional layers of cover lay may be used. In addition to being use for strain relief, stiffeners can be used for component support. Access holes through stiffeners should be from 0.010 to 0.015 in (0.25 to 0.38 mm) larger than the through hole in the flexible printed circuit. Stiffeners can also be designed so as to act as supports that allow the assembly to be wave-soldered.

6.6 Layout Development

Since almost all flexible printed circuits are custom-designed for specific applications, the following information should be developed by the designer:

- Wire list
- Schematic diagram
- Current or voltage drop allowed per conductor
- Capacitance limitations or shield requirements
- Mechanical and environmental data basic to deciding which of the many available production techniques will meet the application at the lowest cost.

Working at any convenient scale, place the various connectors and other components to be wired in their relative position. Next, determine all physical restrictions to the geometry of the circuit. Using the schematic for the circuit to be developed, interconnect the various points. At this time, it will be evident if there is a need for crossovers which may require a multilayer circuit. Except in very complex circuits, it is usually possible to avoid the expense of multilayer construction by arranging the wiring points so that a simple one-for-one wiring plan is achieved.

Begin by reducing the number of layers at points of greatest concentration—usually at an input or output connector. First, find, identify, and list all common terminals. Then, arrange the layers to minimize crossovers, at the area of highest density. Continue crossover reductions to less-dense conductor areas. Thus, the basic principle of fanning out from the areas of greatest conductor density to areas of lesser density provides the best arrangement, both mechanically and at lower cost.

Shielding may affect mechanical flexibility of the circuit, for as the shielding increases, flexibility usually decreases. For applications that require repeated flexing of single- or double-shielded circuits, you must determine where this will occur and the number of bendings required during the operating lifetime. Complex wiring systems or extremely space-limited systems require evaluation with sample patterns (paperdolls) of the flexible circuitry. Prefitting the circuitry avoids possible mistakes in the original design—mistakes arising from the failure to avoid obstructions that you did not visualize in the original layout.

6.7 Design Parameters[1]

6.7.1 Conductor width

The advantage that flat conductors have over round wires in current-carrying capacity is due to their inherently larger ratio of surface to volume. Figure 6.3 is a conductor design nomogram that shows this advantage and can be used in addition to determine flat conductor resistance and AWG wire size with equivalent resistance. The chart

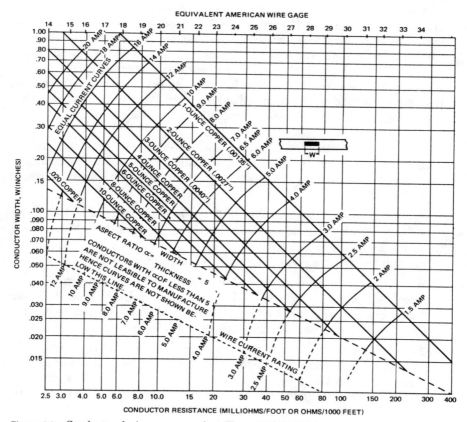

Figure 6.3 Conductor design nomograph. *(From IPC-D-330.)*

may also be used to find the flat conductor dimensions required to replace existing round wires.

While most interconnections carry currents in the milliamp range, there often exist requirements to carry greater currents. Usually, the required circuit is specified as a conductor cross section equivalent to an American Wire Gauge. For example, using Fig. 6.3, assuming a requirement for 28 AWG wire, the flat conductor equivalent for 2-oz copper would be 0.76 mm (0.030 in) in nominal conductor width.

Due to increased area and generally better bond between conductor and insulation than that of round conductors, the flat conductors in flexible printed circuits will dissipate heat more efficiently. If current-carrying capacity is determined by a 30°C temperature rise, it may be possible to up-rate the AWG requirement by one or two numbers.

Figure 6-3 is based on a single flat conductor with 0.010-in (0.25-mm) insulation; however, it should be mentioned that single-layer multiconductor cables will carry approximately half as much current as the single conductor under similar conditions. The values thus derived must be derated if more than one conductor in a cable will be carrying significant current and contributing to the heat load and if the ambient temperature is other than 20°C. A rule of thumb indicates that single-conductor current capabilities should be derated by 20 percent if two conductors will be carrying equal current and by 50 percent if 15 conductors carry equal current. An example of how the nomograph in Fig. 6.3 is used is as follows:

1. *Find the resistance of a 2-oz × 0.030-in-wide conductor:*   Project horizontally from 0.030-in on the width scale to the 2-oz copper curve; then drop down vertically to the resistance baseline. Read 100 mΩ/ft.

2. *Find how much current a 2-oz × 0.030-in conductor will carry for a 10°C temperature rise:*   Project horizontally from 0.030-in on the width curve to the 2-oz copper curve. Read the current from the constant-current curves nearest the intersection.

3. *Find the AWG round wire size with equivalent resistance rating of a 2 oz × 0.030-inch conductor:*   Project horizontally from 0.030 in on the width scale to the 2-oz curve; then project up vertically to the AWG baseline. Read 30 as the closest AWG size.

4. *Find the AWG wire size that can carry a current equal to a 2-oz × 0.030-in conductor:*   Project horizontally from 0.030-in on the width scale to the 2 oz curve. The intersection occurs at the 2-A constant-current curve. Follow that curve down until it intersects the wire current rating Line. Project that intersection vertically to the AWG baseline. Read 28 as the closest AWG wire size.

5. *Find the width of a 2 oz flat conductor that can safely carry 10 A without exceeding a 10°C temperature rise:*   Locate the intersection of the 2-oz curve with the 10-A constant-current curve. Project that intersection horizontally to the width scale. Read 0.31-in.

6. *Find other combinations of width and thickness for conductors capable of carrying 10 A without rising more than 10°C:*   Note the intersection on the 10-A constant-current curve with 3-, 4-, 5-, and 6-oz conductor. Project each of these horizontally to the width scale. Read 0.25, 0.21, 0.185, and 0.165 in, respectively.

7. *Find the flat conductor size for handling 4 A at a maximum drop of 0.125 V/ft:*   Calculate the required resistance per foot from $R = E/I = 0.125/4 = 0.031$ Ω/ft. = mΩ/ft. Locate this resistance on the baseline and project it vertically to intersect the 2-, 3-, and 4-oz curves. Project these intersections horizontally to the width scale. Read 0.098, 0.065, and 0.049 in, respectively. If you were selecting the conductor on the basis of the lowest temperature rise, you would normally choose the 2-oz × 0.098-in size, for it has about 100 percent more surface area than 4-oz conductor and about 33 percent more surface area than the 3-oz conductor.

6.7.2   Conductor spacing

Once a conductor width is chosen, the space between conductors is the next consideration. Often the spacing is chosen by manufacturing considerations rather than by electrical performance requirements. This is because voltages are normally low and most materials used for flexible circuits can withstand 300 V/mil or more. Most circuits feature spaces 0.010 to 0.050 in (0.25 to 0.38 mm). When required, the spaces can be as little as 0.003 in (0.08 mm). Otherwise, the selection of the conductor spacing should be in accordance with the same criteria used with rigid printed circuits.

Since the spatial relationship between conductors in a flexible printed circuit is fixed, it is possible to design the circuitry so that it has uniform electrical characteristics between lines. It is also possible to design in fixed and tightly controlled impedance and capacitance characteristics. The fixed design also provides for a more constant value of cross-talk. For critical digital systems, the rise time, decay time, and propagation velocity can be accurately predicted.

A typical layout can be made by using AWG equivalent for line width. Another approach to layout is that rather than laying out the lines, you are, in effect, laying out the space. Often, this is a simpler way to design and manufacture a circuit. The increase in available copper also makes for a more dimensionally stable circuit and improves the ability of the product to be handled.

6.7.3 Lands

Make lands (terminal areas and pads) as large as possible without violating space requirements. Where applicable, use these minimum limits:

- When the land is at least three times the hole size, specify a minimum of 0.005-in (0.13-mm) copper all around the hole.

- When the land is two times the hole size, copper must be visible all around the hole and a 0.005-in-wide (0.13-mm) land must cover at least 270° around the hole.

- Tie the lands down with anchoring spurs when possible, but avoid creating potential shorts.

- Fillet all sharp bends to reduce stress concentrations.

Since most circuits are terminated in copper lands, consideration should be given to some of the typical hardware shapes and the lands that result from mating with connector pins. It is best to specify pins that are small in cross section so as to minimize the use of circuit real estate. An annular ring around the land must be provided for the connection. Most circuits, however, have one or more lands that cannot be designed this way. In such cases it is necessary to consider alternative land configurations. A possibility would be a land that is odd-shaped and may not have a 360° solder ring.

The use of truncated lands is sometimes necessary. In order to provide maximum adhesion between land and conductor and to provide a solder ring close to 360°, it is desirable to make the land as large as possible. A careful look at the differences between land designs should suggest ways to design acceptable alternate lands. One such technique, to improve solderability and mechanical strength, is to use maximum copper area within circuit geometry and get the most copper around the pin.

6.8 Manufacturing Considerations[2,14]

The methods by which flexible printed wiring is manufactured and the physical limitations of the dielectric used must be taken into account. The method of manufacture will be a function of both quantity requirements where production runs are great enough to warrant the use of punch and die tooling, and where closer end tolerances are at-

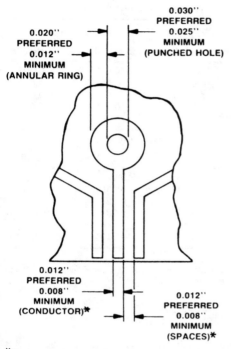

0.030''
PREFERRED
0.025''
MINIMUM
(PUNCHED HOLE)

0.020''
PREFERRED
0.012''
MINIMUM
(ANNULAR RING)

0.012''
PREFERRED
0.008''
MINIMUM
(CONDUCTOR)*

0.012''
PREFERRED
0.008''
MINIMUM
(SPACES)*

Figure 6.4 Recommended feature sizes. *(Courtesy of AT&T Technologies.)*

*Feature sizes less than 0.012'' should be localized and limited.

tainable than in the short-run items which must be fabricated using general-purpose equipment.

The second influence on the choice of the method of manufacture is the type of wiring required. It should be recognized that the minimum number of manufacturing steps will usually result in the lowest-cost product. In general, for optimum producibility the flexible printed circuit features should be designed in accordance with the guidelines of Fig. 6.4 and Table 6.6.

6.8.1 Low-cost design

Some basic factors to achieve low cost design are

- Specify 1- or 2-oz copper conductors. These thicknesses are handled and stocked in large quantities. Other thicknesses are available if design requires, but they may cause delay or increases cost.

- It is recommended that conductor width and spacing be limited to a 0.010-in (0.25-mm) minimum. Smaller line widths and spacing can be fabricated but at increased cost.

TABLE 6.6 Recommended Feature Sizes and Tolerances, Inches*

Design feature	Standard product	Reduced produceability	Tolerances	
			Normal	Minimum
Hole size	> 0.020	< 0.020	± 0.003	± 0.001
Trace width	> 0.010	< 0.010	± 0.002	± 0.001
Space width	> 0.005	< 0.005	± 0.002	± 0.001
Trace to edge of part (steel rule die technology)	> 0.020	< 0.010	± 0.010	± 0.005
		Greatest dimension		
Feature-to-feature location:	≤ 12.0	≤ 18.0	≤ 24.0	
Preferred (MIL-STD-2118)	0.028	0.034	—†	
Class A (IPC-D-249)	0.034	0.040	0.046	
Standard	0.020	0.024	—†	
Class B	0.022	0.024	0.034	
Reduced produceability	0.012	0.016	—†	
Class C	0.012	0.018	0.022	

*1 in = 2.54 cm.
†MIL-STD-2118, in lieu of tolerances, offers the following: "Drawing tolerances must reflect bend and fold allowances between component mounting rigid areas."
SOURCE: AT&T Technologies.

- Design lands oversize.

- Utilize the fewest number of layers both conductor and dielectric.

- Specify enlarged access areas in the covercoat at solder land areas using simple shapes rather than tight-fitting land-sized areas.

- Try to keep punched-out bare conductor areas on same side of the circuit. If it is necessary to have access from opposite, try to avoid the extra cost of punching and laminating by folding the flexible circuit.

- Alignment tolerance between layers should be set a a minimum of ± 0.015 in (0.38 mm); the latter will be preferable from a cost standpoint.

- It will frequently be desirable to include the ground planes in multilayer circuits and these planes will have to have clearance between any holes employed for through connections. The amount of clearance will be dependent upon the current involved, but it is desirable to have a clearance per side of at least 0.030 in (0.76 mm) of any plated-through or eyeletted connection.

6.8.2 Tolerances

Each operation may introduce some error. It should also be remembered that flexible laminates are not as stable as metal, and therefore, it is very diffiucult to hold close tolerances. Liberal tolerances should be used wherever and whenever possible.

6.8.2.1 High quantity. For production runs in which quantities are great enough to warrant the use of piercing and blanking dies, the following rules should be observed:

Hole size. Small hole diameters have been successfully punched, but standard practice recommends 0.020 in (0.5 mm) or larger.

Hole-to-hole and hole-to-edge spacing. The former depends essentially on the tolerances in the construction of the punch and die and the latter on the method used in obtaining the locating holes for blanking. In most cases, material is prepunched and all subsequent operations obtain registration from these holes, thereby holding a tolerance of ± 0.010 in from hole to edge of board. Model lots or short runs of flexible printed circuits present shop problems entirely different from production runs in that expensive punch and dies and special tooling cannot be purchased and still economically produce a small lot.

6.8.2.2 Low quantity. The following rules should be observed for small lot production

- Tolerances of ± 1/32 in (0.8 mm) on shop tools such as bench shears, scissors, hand punching, and drilling of holes by eye.
- Avoid prepunch covercoat and base material as much as possible
- Outside contours, slots, notches, and cutouts should be as simple as possible in the design of short runs

6.8.2.3 All quantities. The following are the recommended minimum tolerances in inches for economical production of flexible printed circuits regardless of lot size.

1. Hole diameter tolerance (see Table 6.7)
2. Center distance between holes
   a. Drilled by eye within ± .0010-in

TABLE 6.7 Hole Diameter Tolerances

Hole diameter, in†	Drilled, in	Punched, in
Up to 1/16 inch diameter	± 0.005	± 0.002
5/64 to 3/16 inch diameter	± 0.005	± 0.002
Over 3/16 inch diameter	± 0.010	± 0.002

*1 in = 2.54 cm.
SOURCE: AT&T Technologies.

   b. Punched with one single-stage punch and die within ± 0.002 in
   c. Punched with two single-stage punch and dies within ± 0.005 in
3. Hole-to-pattern tolerances
   a. Drilled by eye pattern within 0.010-in at center.
   b. Punched with compound punch and die within 0.005-in center.
   c. Registration front to back ± 0.010 in
4. Circuit pattern-to-outside dimension
   a. Sheared edges within ± 0.032 in
   b. Blanked edges within ± 0.010 in
5. Overall dimension tolerance
   a. Sheared edges within ± 0.032 in
   b. Blanked edges within ± 0.010 in
6. Hole-to-outside dimension tolerance
   a. Piercing and blanking die within ± 0.010 in
   b. Drilling and shearing within ± 0.032 in
7. Plating of a minimum thickness only should be specified.
8. Conductor width and spacing tolerances should be as shown in Table 6.8.

These general rules and tolerances should be kept in mind while designing flexible printed circuits for economical manufacturing. It should be remembered that there are many operations, such as laminating, photographing, printing, etching, and plating, that are also necessary to produce flexible printed circuits.

6.9 Cabling Configurations

6.9.1 Right-angle fold

The right-angle fold technique is specifically for a nonmoving application where there is no demand for flex-life but conformance to a given mechanical cabinet. A word of caution: high-vibration environments

TABLE 6.8 Conductor Width and Spacing Tolerances

Weight of copper per sq. ft., oz	Width Tolerance, in*	Spacing Tolerance, in
1	+ 0.002 − 0.002	+ 0.002 − 0.002
2	+ 0.002 − 0.003	+ 0.003 − 0.002
3	+ 0.002 − 0.005	+ 0.005 − 0.002

*1 in = 2.54 cm.
SOURCE: AT&T Technologies.

may dictate that special mounting or other restriction of physical movement be used in the area of the fold.

Flexibility of a given conductor pattern depends on the type of insulation used, conductor dimensions, and other factors. A general rule, however, that the radius of a bend in a one-layer flexible circuit should never be less than (sharper than) three times the overall conductor thickness. For maximum flex-life, it is best to have conductors approach and traverse a fold or bend perpendicular to the fold line.

The folds should be neat and uniform, following the surfaces of the package, in accordance with the bend allowances required in tight corners. Irregular folds should not "bind" against parts, since such binding areas can be points of weakness under vibration. Use a strain-relief bar at points where flexing occurs in a sharp radius bend close to solder joints.

Whenever a cable must assume a fixed curvature or hold, as in a moving cable, the bend may be preformed at the factory, if a formable laminate has been specified. In addition to simplifying installation in an equipment, preformed bends hold a fixed radius of curvature and thus distribute stresses evenly and contribute to longer flex-life and high reliability.

6.9.2 Moving or rolling cable

The prime application for a moving or rolling type of printed circuit is the roll-up which attaches from a sliding drawer to its cabinet. Best results are achieved by a single-layer construction with fully annealed copper conductors on the neutral axis of the roll layer. The choice of dielectric material is limited to attain the window-shade-type configuration to those dielectrics which can be posttreated to create the memory for roll-up. This type of retractable cable has a plastic memory, thus it rolls up on itself when released and returns to its factory-preformed shape, i.e., a cylinder.

Using this technique we have constructed cable assemblies with 170 conductors, which pull out to a 24-in (61-cm) length with less than 5-lb pressure, yet rerolls itself into a compact 2 in (5 cm) in diameter. When considering retractable cables, keep these factors in mind:

- When the drawer is closed, rolled cable can be below or behind the drawer. Allow enough room for the rolled cable.

- Use strain-relief bars on each end of the cable so tension and torsion forces do not reach solder terminations.

6.9.3 Accordion cable

The accordion technique can be used with those dielectric materials which have a weaker spring action. The material is postformed with convolution, which must be properly radiused to avoid a crease and distribute stress gradually over the bending area. This type of wiring usually requires some form of external restriction member to control its position in the retracted condition.

References

1. "Flexible Printed Wiring," *Printed Wiring Design Guide*, IPC-D-330, Section 6, Institute for Interconnecting and Packaging Electronic Circuits (IPC), Lincolnwood, Ill.
2. Joseph C. Fjelstad, Printed Circuit Builders Inc., "Design Guidelines for Flexible Circuits," *Electronic Packaging & Production*, August 1988, pp. 32–34.
3. "Design Standard for Flexible Single- and Double-Sided Printed Boards," ANSI/IPC-D-249, January 1987, Institute for Interconnecting and Packaging Electronic Circuits, Lincolnwood, Ill.
4. "Flexible and Rigid-Flex Printed Wiring for Electronic Equipment, Design Requirements for," MIL-STD-2118, May 1984.
5. David Becker, Sheldahl Corporation, "Flex Circuitry: Designing for Manufacturability," *Printed Circuit Design*, August 1988, pp. 54–56.
6. "Specification for Single-Sided Flexible Printed Wiring," IPC-FC-240, January 1979, Institute for Interconnecting and Packaging Electronic Circuits, Lincolnwood, Ill.
7. Sheldahl Electronic Products Division, "Basic Guidelines for Flexible Printed Circuitry," Northfield, Minn.
8. Robert F. Currier, Lockheed Missiles & Space Company, "Problems in the Production of Rigid-Flex Multilayer Printed Circuit Boards," IPC-TP-176, IPC Technical Paper, Institute for Interconnecting and Packaging Electronic Circuits, Lincolnwood, Ill., September 1977.
9. "Performance Specification for Rigid-Flex Printed Boards," ANSI/IPC-RF-245, April 1987, Institute for Interconnecting and Packaging Electronic Circuits, Lincolnwood, Ill.
10. "Flexible Bare Dielectrics for Use in Flexible Printed Wiring," ANSI/IPC-FC-231, February 1986, Institute for Interconnecting and Packaging Electronic Circuits, Lincolnwood, Ill.
11. "Flexible Metal-Clad Dielectrics for Use in Fabrication of Flexible Printed Wiring," ANSI/IPC-FC-241, February 1986, Institute for Interconnecting and Packaging Electronic Circuits, Lincolnwood, Ill.
12. "Flexible Adhesive Bonding Films," ANSI/IPC-FC-233, February 1986, Institute for Interconnecting and Packaging Electronic Circuits, Lincolnwood, Ill.
13. "Adhesive Coated Dielectric Films for Use as Cover Sheets for Flexible Printed Wiring," ANSI/IPC-FC-232, February 1986, Institute for Interconnecting and Packaging Electronic Circuits, Lincolnwood, Ill.
14. AT&T Technologies, Berkeley, Heights, N.J., "Flexible Printed Wiring Design Guide," MN85-05PWB, 1985.

7

Computer-Aided Design Setup

7.1 Introduction[1]

There are three key reasons to use computer-aided design (CAD) tools to design printed circuit boards. First is increased productivity. Increased demands on product development cycles require increases in productivity that are beyond human limits without CAD tools. With CAD, designs can be edited faster—actually automating the engineering change order process. You will no longer be required to do many of the tedious mundane error-checking processes manually; your product documentation quality will likely be much higher, and you will probably find that CAD helps to reinforce the use of design process metrics to keep the product schedule tightly managed and in control.

Second, reason is increased design complexity. Increases in the complexity of technology may force the use of CAD tools. Even if the design is not already challenged by the following product criteria, they are probably within the planning range and can dictate the need for printed circuit board CAD because of:

- Tighter manufacturing tolerances
- Higher yield rates
- Higher packing densities
- Smaller components
- Smaller line widths and spaces
- Newer types of devices and technologies
- Lower manufacturing costs

CAD can deal with all of these issues simultaneously in a controlled manner.

Third reason is increased integration requirements. As technology demands increase for the design team, so do the corresponding requirements for all departments. Therefore, it is important to take extra steps toward increased integration among these departments to ensure the integration of simulation, management resource planning, documentation outputs, and manufacturing. CAD also helps in automating and standardizing design and manufacturing databases.

Quality is improved when a better product is produced in less time and with fewer risks. CAD tools impose discipline and design techniques on the designer which will help bring consistency to the design process. Once started using CAD tools, it is possible to check the designs more frequently and catch errors before they become expensive mistakes; i.e., no more boards produced with conductors missing or output lines that never connect to edge connectors.

In addition, printed circuit board CAD packages that are designed to take advantage of more sophisticated manufacturing techniques like via minimization and diagonal routing will help improve final product quality.

7.2 Design Methodology[2]

The impact on an existing CAD system, or on the process of selecting a CAD system, is significant if the user is considering the introduction of a new technology into a manufacturing or engineering program. Several factors need to be considered as an integral part of the overall design guidelines when establishing a design method. No one system can address al the technological issues presented to the designer, and the ability to push a design through CAD systems often depends on the ingenuity of the system operator and the circuit designer. The optimum CAD system has excellent features for solving today's design requirements, plus the flexibility to allow human intervention.

The most pressing problems faced by CAD system developers and operators today are board size and density. With most systems, the designer can find ways to work around these problems by fooling the system, changing the grid, or splitting the board. Also, higher-technology designs involving extremely dense analog circuitry typical of high-speed RF circuits, fine-line technology, surface-mount technology, and thick-film design and flex circuitry pose a considerable challenge to CAD system developers and to the designers that must work with existing CAD systems.

7.3 Input Packages[2,3]

Design and documentation feature requirements that apply to the basic layout, the printed wiring master artwork, the printed circuit board itself, and the end-item printed board assembly, all must be taken into consideration during the design of the board. Therefore, proper data must be supplied to the printed circuit board designer (Fig. 7.1). When component process and producibility issues have been resolved, substrate circuit layout can commence. This involves cou-

Title_____ Entry Rev. No._____ Designer_____ Ext._____

ML No._____ Project_____ Work Pkg. No._____

Job No._____ SS No._____ POE Req'd ☐ Yes ☐ No

BOARD DESCRIPTION

Layout Scale_____ Min. Grid Size_____ Schematic Input   ☐ Yes ☐ No

Multi-Use Board   ☐ Yes ☐ No   How many more to come_____

Feed thru size   ☐ .036 Standard   ☐ .022 High Density

Board   ☐ Two Sided   ☐ Multilayer   ☐ Other   Highest voltage_____

Foil thickness (oz.) Surface_____   Other layers_____

| Hole | ☐ Yes | Silkscreen | ☐ Yes | Marking | ☐ Yes | Soldermask | ☐ Yes |
| Master | ☐ No | Master | ☐ No | Master | ☐ No | Master | ☐ No |

CHECK TO VERIFY DESIGN INPUT IS STANDARD AS FOLLOWS:

1. ☐ Crop marks are on all corners of all layout layers.
2. ☐ All mounting/hardware holes not on grid or pattern are dimensioned from board datum.
3. ☐ Parts list is filled out or all components are called out on layout.
4. ☐ The gerber aperture is called out for all line widths — not the line width desired on the finished board.
5. ☐ Legends are clearly shown on mylar drawings.
6. ☐ All pad apertures and hole size callouts are on the layout.
7. ☐ Oblong pads desired are not laid out as round pads on layout.
8. ☐ All pads on .100 centers with a trace between pads call out a .012 gerber aperture (.008 min. conductor callout on drawing.)
9. ☐ All design lines and spacings have been expanded where space permits per Standards memo of 8 February 1982.
10. ☐ Silkscreen text is visibly located — not under component bodies, heat sinks or overlay features.
11. ☐ "Keep out" areas are defined. No pads, circuitry, hardware allowed in area.
12. ☐ All conductive pads and traces are .020 min. from protected board edges, cutouts, holes.
13. ☐ No silkscreening is designed over traces on foil greater than 1 oz.
14. ☐ All unprotected board edges have .100 clearance from conductive patterns.

Comment: _____

_____

_____

Note: Designer fills out this form, files the white copy and attaches the pink copy to the drawing

Figure 7.1 Printed circuit board CAD input form. *(Courtesy of Singer Librascope.)*

pling component size, pinout, performance, and testability requirements with established design rules.

The inherent electronic performance improvement, resulting from reducing the magnitude and complexity of interconnections, must be balanced against possible interference due to increased packaging density. With experience, design rules will evolve which increase the predictability of this trade-off, but verification is best accomplished by prototype evaluation. Additional trade-off considerations include performance criteria, cost projections, producibility, and the status of manufacturing.

7.3.1 Schematics and logic diagrams[1]

The initial schematic or logic diagram designates the electrical functions and interconnectivity of the printed circuit board and its assembly. The schematic diagram should define component identifier, logical connectivity, and, when applicable, critical circuit layout areas, current requirements, shielding requirements, grounding and power distribution, the allocation of test points, and any preassigned input/output connector locations. Schematic information may be hard copy or either manually or automatically generated computer data.

Most CAD packages offer several methods of entering necessary information into the system. Normally, information is entered through a schematic capture package, which may or may not be totally integrated with the printed circuit board design portion of the system. Schematic capture and printed circuit board design could be developed by two different companies and then linked together with an interface program developed by either one of the companies, or as a combined effort of both. Both the schematic capture package and the printed circuit board design package can be developed by the same company allowing integration and the forward and backward transfer of information between the two.

Another method of entering the necessary information into a printed circuit board design package is through an ASCII net list. This method allows the user to compile the schematic information into a text file of a prescribed format and then process this ASCII net list through a utility program which will convert the net list information into a printed circuit board database, which can then be read and manipulated by the printed circuit board design package. The input of a net list into the CAD system represents the fastest method of giving connectivity information to your CAD system. However, it does not provide a pictorial schematic diagram for drawing release purposes. Net lists can be hand-entered, down-loaded, or extracted from a CAD schematic generation system or engineering workstation.

7.3.2 Parts list

The parts list may be hand-written, typed on a standard form, computer-generated, or written on the schematic. When selecting the parts, you should consider the applicability of the parts for printed circuit board mounting, lead lengths and diameters, through holes, surface mount, standard mounting grid, hold-downs, adhesives/chips, etc. The initial parts list must contain the component identifier, quantity of each part, and applicable part number.

The initial parts list may also contain a complete description of the part, quantities, manufacturer, and special ordering instructions. Sometimes, for cost analysis purposes, it is also desirable to include the cost of the items needed. What information must be included and how it should be formatted depends upon the standards established by each user.

The parts list is on a separate sheet or in a separate computer file; it must have some identifier which relates it to the schematic. A good practice is to include the name of the board for which the parts list was prepared, the drawing number, including the revision, and the date with the names of the persons preparing, checking, and approving this document. Some means of identifying the revision level must also be provided.

When nonstandard parts are used in the design, the detailed electrical and mechanical/physical description and manufacturer data sheet should be provided. Reference this information to the schematic diagram so that the appropriate layout configuration can be determined. Because nonstandard parts are used and only manufacturing data are available, a specification control drawing should be developed to ensure that future parts match the parts selected for this design. The specification control drawing should provide sufficient details to completely describe the part, including the suggested vendor's part number.

7.3.3 End-product requirements

The success or failure of a board design depends on many interrelated considerations. For example, the following parameters may affect end-product usage, so the designer should consider them:

- Equipment environmental conditions, such as ambient temperature, heat generated by the components, cooling ventilation, and vibration.

- Maintenance philosophy during the service life of the equipment, especially with respect to component placement that affects component accessibility (components one-side/components two-side).

- Spacing between boards that might limit lead protrusions and affect the placement of brackets and hardware.

- Testing and fault location requirements that might affect component placement, one-sided or two-sided conductor routing, connector contact allocation, etc.

- Manufacturability-level requirements (quantities) must be considered, i.e., products that will require large quantities of printed circuit board's must be producible at a high production level.

- If an assembly is to be repairable, you must consider component and circuit density and the selection of board and conformal coating materials. In general, repair criteria should adhere to the guidelines of IPC-R-700.[4]

- Special requirements such as the charts, tables, and illustrations found in the IPC design standards, such as IPC-D-319 and IPC-D-949, to help guide you through your design.[5,6]

7.3.4 Testing

Good design practice dictates that prior to starting a design, a testability review should be held. Testability concerns, such as test access, circuit visibility, circuit operation, and special test requirements and specifications, are discussed as part of the test strategy. Any conductive pattern board configuration concepts necessary to test the board shall be incorporated into the final printed circuit board layout.

7.3.5 Board mechanical outline[7]

The board outline requirements affect good board design. The following should be considered by the designer:

- Electrical restrictions such as the number of layers and layer distribution and material selection criteria.

- Mechanical restrictions such as connector locations, cutouts, and clearances around the board.

- Component mounting and soldering technologies [wave solder, vapor phase, infrared (IR), etc.] to be used on final board.

- Where and how the board is mounted in the next assembly and how it is electrically connected to the assembly. When selecting the location of interconnect points, consider proper flow on and off the board to minimize the effects of electrostatic discharge (ESD) and electromagnetic interference (EMI).

- When two assemblies are mounted face to face, component placement considerations relative to component height and interferences must be considered.

- The board outline should be defined in relation to an internal datum. The board outline should be on-grid with the datum located in the lower left corner of the board. Keep cutouts and notches to a minimum. Because both metric- and inch-based grids are used, the difference between soft (exact) conversion and hard (approximate) conversion can make a difference in the board profile. The design should have one or the other grid reference, not both.

7.4 Job Setup[2]

Before a printed circuit board can be designed, certain tasks must be fulfilled in a specific order so that the design process can proceed in a smooth, orderly manner. In order to take full advantage of the CAD capabilities, it is necessary to establish specific rules which the computer will follow. Applicable design standards should be followed in establishing the design rules.

7.4.1 Modular grid

The choice of the modular grid system used in the design of printed circuit board is based on the dimensioning system used on the majority of the component terminal locations of parts mounted to the printed circuit board. For designs where the majority of component land locations are metric- (SI-) based, the basic modular units of length are 2.0, 1.0, 0.5 mm or other multiples of 0.5 mm and should be applied in the X and Y axes of the cartesian coordinate system. For designs where the majority of component land locations are inch-based, the basic modular units of length are 0.100, 0.050, and 0.025 in, or other multiples of 0.005 in, and shall be applied in the X and Y axes of the cartesian coordinate system.

7.4.2 Conductors dimensions

The minimum width and thickness of conductors on the finished board should be determined on the basis of the current-carrying capacity required, and the maximum permissible conductor temperature rise and etch factor. Typical design aids can be found in the IPC design standards IPC-D-319[5] and IPC-D-949.[6] Wherever possible, the design should incorporate processing allowances which will allow the manufacturer to produce a part that will meet the end-product requirements. For some CAD systems, a good practice is to keep the number of different conductor widths to a minimum to facilitate conductor routing.

7.4.3 Land-hole relationships

Good practice is to establish a standard which contains component-hole-to-land relationships and holes sizes for a variety of lead diameters. When determining the land to hole size relationship, the design should meet land to plated through hole requirements of documents such as IPC-D-275.[8] Minimum annular ring requirements, the end-product hole characteristics, and the processing allowances should be established in accordance with both design standards and any automatic insertion hole requirements. (See Chap. 5.)

In addition, minimum hole-to-hole and hole-to-board edge requirements should be established. If a mechanical photoplotter is used, these same standards may be applied to laser plotters. However, there is more flexibility since laser plotter limitations are related to laser spot size, not the number of apertures held in any aperture wheel (see IPC-D-310).[9] A photoplotter aperture list should be established based on the user standard component types.

7.4.4 Conductor spacing

Minimum conductor spacing should be established for the following conditions:

- Between different voltage-carrying conductors
- Conductor to land areas
- Conductor to board edges
- Conductor to mounting holes
- Conductor to other metal objects located on the board or objects used to mount the printed circuit board to the assembly

7.4.5 Component mounting

General rules for component mounting and spacing should be established. See IPC-CM-770,[10] IPC-CM-780,[11] and IPC-SM-782.[12] Typical considerations should include component distance from board edges, spacings between components, component distance from mounting holes, and specific component orientations which affect soldering and/or automatic insertions.

7.4.6 Large conductive areas

The criteria for determining a large conductive area should be specified. For example, any area that extends beyond a 1-in-diameter (2.54-

cm) circle should contain etched clearance areas that will break up the large conductive area but retain the continuity and functionality of the conductor, such as with a cross-hatched ground plane. Large conductive areas should, if possible, be on the component side of the board.

Electrical considerations may affect the design of large conductive areas and may require additional vias and tie planes on alternate layers. Large plane areas may need to be unbroken at certain frequencies and voltages. In addition, thermal relief criteria should be specified for lead attachment to large conductive areas. Thermal relief criteria, such as the width of the conductive tie, the width of the clearance, the land size, and the direction of the conductive tie, should be specified.

7.4.7 Orientation symbols

Special component orientation symbols may be incorporated into the design to allow for ease of inspection and assembly of the parts. Techniques may include special symbols or special land configurations for ICs and surface-mount devices. Examples of this technique are square lands or dots denoting terminal no. 1 of ICs, plus signs for polarized capacitor anodes, and numbers and letters for connector pin identification.

7.4.8 Solder masks

Solder mask coatings are recommended to prevent solder bridging between conductors during the soldering process as well as enhance electrical performance. Various materials can be used as coatings, each having specific characteristics which affect the mask to land area clearance requirements. (See IPC-SM-840.[13]) Printed circuit board assembly technologies and board complexity determine the minimum solder mask clearances recommended for the land areas used in the design. Different solder mask materials also have differing clearance requirements.

7.4.9 Conformal coatings

Conformal coatings are used for environmental protection as well as mechanical support for components mounted on a printed board. The designer should incorporate any end-product requirements for conformal coating so as to design for proper component mounting as well as circuit routing. Conductor spacing may vary, as may the number of conductors routed under a component or the accessibility to via sites under the component.

7.5 Libraries[2]

The libraries available with CAD packages are usually determined by the sophistication of the package. A minimum library set should, in sufficient variety, address 80 to 90 percent of design requirements. How a library is created, maintained, and used is critical to the ease in which a printed circuit board can be designed and how easy it will be to read and to alter.

For example, libraries should be able to handle multiple land definitions to allow for a surface-mount component's land pattern. In addition, component land patterns must supplement companion geometries that describe the solder mask pattern and the solder paste stencil used in conjunction with each component. Some approaches use the same image to reflect all their requirements and then let the manufacturer modify the phototool to provide the necessary relationships. Thus, the CAD library only needs to contain one land pattern definition. It is also important to have mirroring capability to allow for components on the "other" side of the board. Presently, when using systems that do not adequately handle these library requirements, the designer must keep detailed documentation updated to reflect what the CAD system does not recognize and check.

CAD systems require libraries of component shapes. These shapes represent both the electrical function and mechanical size and shape of various components. Depending upon the CAD system used, a relationship between the electronic schematic symbol and the mechanical part shape may or may not be required or available. If a relationship between schematic and part shapes is required, pin assignments and sequence must be consistent between the symbols and the part shapes because of net list extraction. A carefully coordinated effort makes design data transfer easier between schematic and board layout and board layout to computer-aided manufacturing (CAM) postprocessing.

Some systems require the user to have a physical package with each logic, discrete, and connector element. Other systems use pointers that locate a physical package in the database without storing the graphics. Packaging should be flexible enough to allow the same element in different case types.

Physical shapes can be classified into at least two categories. One depicts the outer perimeter of the component body, and the other depicts the actual part configuration, (Fig. 7.2).[7] Library figures (component outlines) are usually drawn to the maximum size of the component and lands rather than nominal or minimum sizes, see IPC-D-319[5] and IPC-D-949.[6] When establishing the physical shape, consider clearances for automatic insertion equipment and other factors that would affect component placement (see IPC-CM-770[10]).

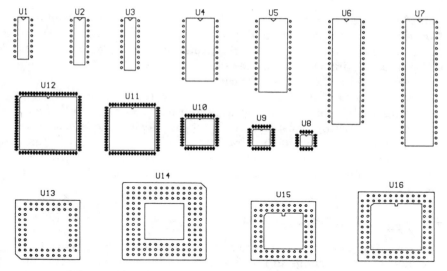

Figure 7.2   Library parts.   *(Courtesy of Personal CAD Systems.)*

Some CAD systems require a data table that shows the relationship between the physical shape of a component and its electrical symbol. Information such as interchangeable gates within a package, between packages, and swappable pins are examples of this kind of data (see IPC-D-352[14]).

How a library is created, maintained, and used will be critical to the ease in which a circuit is drawn and how easily it is read and altered. If the system offers many different libraries, check to see how complete they are and what standard is used. Most designers use a variety of technologies when creating their circuit; be certain your design requirements can be met. If the basic system does not include libraries, they could be a costly option.[15]

7.6 Schematic Capture[1,2]

The creation of a schematic diagram on the CAD system will give the user a releasable drawing. Some CAD systems will be able to extract the connectivity information from the schematic diagram and then compare the finished board connections to the schematic diagram or the extracted net list.

Critical pathing and critical grouping capabilities are available in some schematic capture programs. Components can be added or deleted from a critical group or path and the ordering of components in a critical path can be modified as required. The system provides for relocating all components identified in a critical path or group to be moved with a simple MOVE command, along with their associated nets. Notation of critical pathing and grouping can be displayed or not displayed based on hard-copy media requirements. Once identified on the schematic, this information is transferred to the automatic placement program to be used as an aid for component placement.

There are a variety of simulation tools available for all stages of designing: ASIC, analog and digital, and a new technique called electronic breadboarding. Most schematic capture design packages can interface to at least a few of the simulation tools via a net list. The integration may allow for back annotation, input of data in a graphical mode, or even interaction between the schematic capture and simulator.

7.7 Return on Investment (ROI)[1]

There are three major tiers in the market of printed circuit board CAD systems. The price range in the first tier is $500 to $5000; the second tier range is $5000 to $25,000, and the third tier is $25,000 and up per user. But the price can continue to increase if you make a wrong choice when purchasing your CAD system. Statistics show that as many as 20 percent of CAD owners replace their systems for a new one. Because of this it is desirable to seek out as much information about the candidate system as possible to avoid wasting thousands of dollars. The dollars lost are not only in the CAD system replacement and retraining but in recovering full designs in transferring data from the obsolete system to the new system.

7.7.1 Hardware costs

Printed circuit board CAD hardware also ranges greatly in price. You may be able to spend as little as $2000 for a bare-bones clone, low-end design platform, but the price tag can go up quickly with the addition of advanced graphic packages, larger disk drives, faster central processing units (CPUs), more memory, and a variety of peripheral devices. The real hidden costs to be aware of are requirements that will force the hardware configuration to become dedicated to printed circuit board CAD design only. This may be all right if the volume of work supports the investment. However, it is important to consider first if there are any other CAD tools, management tools, or general word processing requirements that should also be supported by the same equipment.

7.7.2 Ongoing costs

Basic ongoing costs for maintaining a CAD system will include any labor and parts required for maintaining the hardware, and update costs for maintaining the total system as new enhancements are developed. Maintenance contracts are generally available for both hardware and software, which are considerably less expensive than buying by the piece or paying by the hour.

Some CAD users want to add in the savings of receiving updates automatically rather than having to go through justifications for each step, then ordering and waiting for the work to be processed. Another point worth evaluating is that hardware costs are not only greatly reduced when working with personal computer–based equipment but maintenance on personal computers is minimal compared to that of workstations.

7.7.3 Additional cost considerations

Additional cost questions which should be taken into account when purchasing a CAD system include:

- What kind of cost will the user pay if the system being considered is outgrown? Is an upgrade path available from the current vendor or will the user be looking at a completely new system from an alternate vendor? Are the databases compatible?

- Are the needed interfaces available and at what price? Some possible requirements are linking or fabrication to manufacturing process; does the vendor support an open architecture to develop custom links?

- How complete are the libraries for the user's design needs? It can take an average of 4 hours each to properly create, check, and document new symbols and parts, unnecessary task with the right system.

- Ideally, key users should be trained on the operation of the system. Thus, it is advisable to allow for the cost of training for system updates. Be sure to consider the cost of the time off the job for trainees and any potential turnover which would require the training of new users. If it is planned to train a user who will then train others, don't forget to allow for their time preparing classes, teaching, and supporting follow-up questions.

- The availability of service bureaus will also be key because they will be able to pick up occasional overflow work. Therefore, the CAD user should find out where they are and whether they are fluent with the CAD system and version. It is not recommended to pay to train a service organization, and do not be disappointed that the designs worked on outside are no longer compatible with the internal CAD system.

7.7.4 Lease options

Lease options are sometimes available. The full cost of a complete system can be placed on a lease. A lease on your printed circuit board CAD system offers you good financial benefits—the main one being that you do not have to pay up front for the entire system, allowing you to use the system productively so that you are getting paid back from its use as you use it. A typical lease can include:

- Software
- Maintenance contract
- Computer hardware and peripherals
- Shipping costs
- Installation and training
- Taxes

7.7.5 Example

System return will vary, but the example analysis shown in Tables 7.1 to 7.3 provides a scenario using typical CAD tools that show how the return on investment (ROI) will be significant using CAD over a manual printed circuit board design system.

Note. This example assumes that the necessary hardware is already owned. If it is necessary to spend an additional $4000 to $15,000 on hardware, the payback period would still be less than 6 months.

As can be seen, the payback from investing in a CAD system is significant, and larger designs generally will result in even faster payback because of the additional savings when using a typical CAD design rules check feature and when editing on the screen vs. manually. Also, these are first-year savings and payback calculations.

7.8 Interface Techniques[2,16]

While logic and circuit simulators, layout editors, tester specter generators, and placement and routing tools are all considered part of the automation process, they remain, by and large, individual islands of progress. Standard data formats that will allow data to pass between these pieces are only now being developed, and tools to manage and integrate the volumes of data remain largely neglected.

The primary problem facing the electronic design community is the tremendous amount of documentation that must be created during the design process. The solution is to streamline the design and documen-

TABLE 7.1   Small (75-Component) Printed Circuit Design Example

Effort	Manual hours	CAD hours
1. 2 "D" size schematic sheets	8	4
2. Checking (assumes drafter inputs schematic)	2	1
3. Make changes	2	1
4. Placement	8	1
5. Routing (assumes 95% route)	24	1
6. Editing*	0	1
7. Checking	8	1
8. Taping or digitizing	16	0
9. Artwork	(1–4 hours)	
Total time to create prototype (film):	70	15

*Manual editing time is included in routing times.
SOURCE: Personal CAD Systems.

TABLE 7.2   Design Example Payback Analysis Data and Assumptions

Expense	Magnitude
1. Hourly rate including fringe benefits	$24/hr
2. Labor hours/year spent on board design (assuming that the designer worked one shift and actually spent 75% of time designing boards)	1500 hr
3. Approximate annual cost for designer time	$36K
4. Labor hours to design 75-component board manually	70 hr
5. Labor hours to design 75-component board	15 hr
6. Software cost estimate (including full maintenance for first year)	$6.61K

SOURCE: Personal CAD Systems.

TABLE 7.3   Design Example Payback Analysis Calculations

Item	Results
1. Typical number of manual designs in one year	12 boards
2. Typical number of designs in one year	25 boards
3. Cost per board if manually designed ($36K/year, 12 boards/year = $3000/board)	$3000/board
4. Cost per board ($36K/year, 25 designs/year = $1440/board)	$1440/board
5. Savings per board ($3000 − $1440 = $1560/board)	$1560/board
6. Number of designs to pay back CAD system cost in 1st year ($6.6K system cost, $1560/board = 4.23 boards)	4.23/boards
7. Approximate payback time in months (4.23 boards, 25 boards/year = 0.169 years = 2.03 months)	2 months

SOURCE: Personal CAD Systems.

tation of advanced digital systems. Existing hardware description languages are not capable of accomplishing this because their evolution was not planned. The need for a hardware description language to input design data became very apparent by 1979. This language would

- Provide a human-readable design description.
- Support the communication of design data among vendors (second sourcing) and between vendors and users (design specifications).
- Integrate the activities of designers working at different levels of abstraction.
- Increase the reusability of hardware designs and descriptions.

In addition to a hardware description language, an intermediate format is required to interface between standards and languages. Thus, several standard languages already have been developed that can be used to describe various automation functions, including printed circuit boards, phototooling, or information related to these topics. Some of these languages have not necessarily been developed for printed circuit board functions, but because of their applicability to other facets of the design process, they can also be used in printed circuit board applications. These include such standards as:

- IPC-D-35X
- IGES/PDES
- VHDL
- EDIF

Their interrelationship is shown in Figs. 7.3 and 7.4. In addition, users may have their own native data format standards. These are sometimes related to the vendor equipment purchased by the user.

Some companies have standardized within the various departments in their organization in order to allow for electronic communication of product data. These are usually proprietary formats and not industry standards. Such digital descriptions are desired in order to facilitate the automation process for producing parts. On-line electronic data transfer is possible and intended to

- Eliminate the need for human intervention.
- Facilitate the storage of documentation in a format other than paper (archiving).
- Create a standard means of describing design data that can be transported to machines other than the one on which they were created.

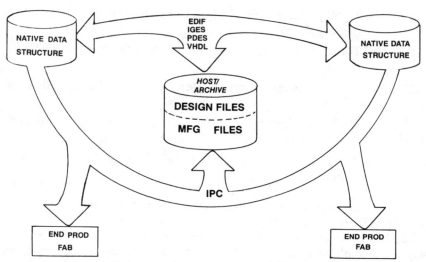

Figure 7.3   Electrical data structure interfaces.   *(From IPC-D-390A.)*

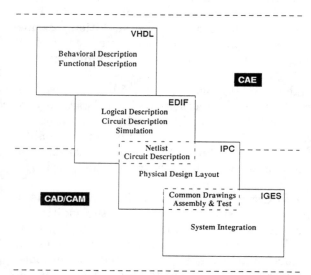

Figure 7.4   VHDL, EDIF, IPC, and IGES relationships.

7.8.1   IPC-D-35X series[17–19]

The IPC has generated a set of companion documents, the IPC-D-35X series, that consists of

- ANSI/IPC-350, "Printed Board Description in Numeric Form"[20]
- ANSI/IPC-D-351, "Printed Board Drawings in Digital Form"[21]

- ANSI/IPC-D-352, "Electronic Design Data Description for Printed Boards in Digital Form"

- ANSI/IPC-D-353, "Automatic Test Information Description in Digital Form"

- ANSI/IPC-D-354, "Library Format Description for Printed Boards in Numeric Form"[22]

- IPC-DG-358, "Guide for Digital Descriptions of Printed Board and Phototool Usage per IPC-D-350"

- ANSI/IPC-NC-349, "Computer Numerical Control Formatting for Drillers and Routers"

The IPC-D-35X series of standards was written for the purpose of digitally describing the logical and physical elements necessary as input to a design system. A part of this description is the network or interconnection of physical and electrical parameters between the various electronic parts. The IPC-D-35X series of standards was also developed to specify printed board data in a machine-independent digital format for communication from design to production (Fig. 7.5). This series has evolved to encompass fabrication, documentation, assembly, and testing. The format supports data communication among computer-aided engineering (CAE), design (CAD), and manufacturing (CAM) systems.

The IPC series of standards started with IPC-D-350 that is used to describe the fabrication and artwork of a printed circuit board. It has grown into a series of standards that describe the printed circuit board in many different ways, including the schematic diagram, assembly drawing, electrical description, test data, etc. These standards also have the potential of describing the systems in which the boards reside.

The use of the IPC-D-35X series is flexible as it allows the user many options. The structure of the standards allows for the addition of new information and new concepts. This is done by developing new data information modules that describe a particular parameter or facet of the design. As future enhancements are required, the concepts used in the IPC-D-35X series can have record formats added to assist in describing parameters needed by the design community.

7.8.1.1 Data language manipulation. The IPC-D-35X series has been developed with the capability for identifying a change in language in midstream. Thus, in a particular data file for a specific printed circuit board, the information might start out in a IPC-D-350 format, switch to IGES, switch back to IPC-D-350, switch to EDIF, and finally con-

All Operations Codes have three digits. The use of the decimal-point symbol (.) in the listing below indicates the position of a digit and illustrates that the Operations Code for these records also may be interpreted as three single-digit Operations Codes in adjacent positions.

GENERAL RECORDS
Columns
123

000 or 0 . . or . 0 . or . . 0 Continuation in present mode; valid for all types of records.
999 End of job.

LINE RECORDS
Columns
123
1 . . Begin new line containing one or more line segments.
. 1 . Linear interpolation.
. 2 . Circular interpolation.
. 4 . Linear "paint in" area outline.
. 5 . Circular "paint in" area outline.
*. 7 . Linear part outline.
*. 8 . Circular part ouline.
. . 1 Format #1
. . 2 Format #2
. . 3 Format #3
. . 4 Format #4

SUBROUTINE DEFINITION RECORDS
Columns
123
2 . . Begin subroutine definition
. 1 . Subroutine definition of a complex feature (executed by a subroutine CALL)
. 2 . Definition of a Point Record special shape (executed as a point record with generic code of G4)
. 3 . Subroutine for dimension description or call out with arrow head
. . 1 Format #1
. . 2 Format #2
. . 3 Format #3
. . 4 Format #4
299 End of subroutine

POINT RECORDS
Columns
123
3 . . Begin new point record.
*. 1 . Feature (land) and hole concentric at the point.
. 2 . Feature (land) only at the point.
*. 3 . Hole only at the point.
*. 4 . Tooling feature and hole at the point.
. 5 . Tooling feature only at the point.
*. 6 . Tooling hole only at the point.
. . 1 Format #1
. . 2 Format #2
. . 3 Format #3
. . 4 Format #4

SUBROUTINE CALL RECORDS
Columns
123
4 . . Subroutine call
. 1 . Linear repeat or step-and-repeat
. 2 . Rotary repeat or step-and-repeat
. . 1 Format #1[1]
. . 2 Format #2[1]
. . 3 Format #3
. . 4 Format #4

ANNOTATION RECORDS
Columns
123
5 . . Begin annotation/dimension record
. 1 . Begin new annotation record
. 2 . Subroutine call for dimension guidelines, call out guidelines, and the appropriate annotation records to go along with that subroutine.
. . 1 Format #1[2]
. . 2 Format #2[2]
. . 3 Format #3
. . 4 Format #4
000 Columns 4-72 contain the text. This record must follow 51.

Format #1: Positional data (LDA Columns 31-72) formatted as three adjacent X and Y fields.
Format #2: Positional data (LDA Columns 31-72) formatted as two adjacent X and Y fields.
Format #3: Positional data (LDA Columns 31-72) formatted as two adjacent XYZ fields.
Format #4: Positional data (LDA Columns 31-72) formatted as one set of XYZ fields. The second set of XYZ fields must be on the following continuation record.
1. The first X, Y data points define the position of the subroutine pattern datum point, or origin. The second and third X, Y data points are used for linear or rotary replications arguments. The angle is specified in degrees for U.S. units and radians for metric units.
2. The first field contains the XY coordinate for the point of the arrow head, if only one arrow head is present. The second field contains the X and Y coordinate for either the line end or the center of a circle subscribing the call out designator. If two arrow heads are present, the first X and Y coordinate will be for the point of the arrow head closest to datum 00. The third field is left blank.

*NOTE: Asterisks only for DIM A or DIM B data subsets (see IPC-D-350 Section 4.1.3.)

Figure 7.5 Operations codes descriptions. *(From IPC-D-350C.)*

clude in IPC-D-350. This manipulation of data languages is intended to enhance the usability of IPC-D-350 in relationship to the other languages. An example of how such an engineering and design system could be configured is shown in Fig. 7.6. Also illustrated are the types of application tools for which the data standards are intended. Presently, other languages do not easily identify the return to IPC-D-350,

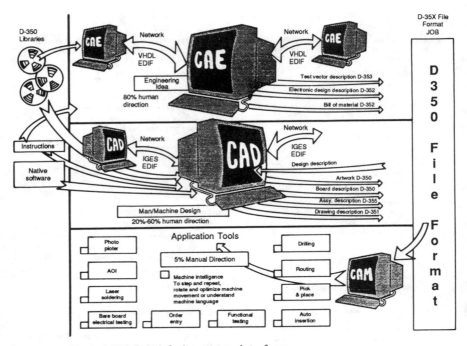

Figure 7.6 Typical IPC-D-350 design system interfaces.

if that is required. This must be checked by the user in order to make a determination on how to move from other languages back to the IPC-D-350 format.

7.8.1.2 Levels of implementation. Since the IPC-D-35X series has varying levels of complexity, a clear understanding of the ramifications of the various formats, conventions, and techniques is important for the user, the developers of translators, and individuals receiving IPC-D-35X data for further postprocessing. It is also important to understand the levels of complexity in order to make certain that data generated at one site can be understood at another site. Three levels of implementation have been identified that reflect on the complexity of computer systems and translators used to convert the standard language to processible data (Fig. 7.7).

Level 1. Level 1 implementation includes the use of line, point, annotation, and subroutine records to represent the image of the end product, e.g., printed circuit boards and artwork. This level is very similar, but has additional intelligence, to that of the language used to define control data for vector plotters. Lines represent stroke data,

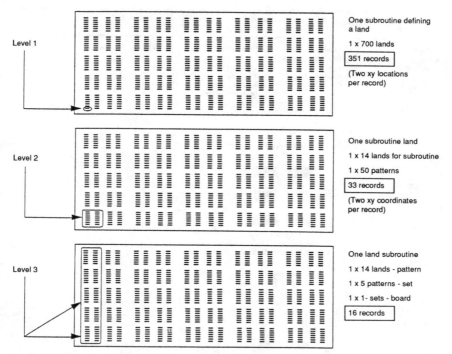

One subroutine defining
a land

1 x 700 lands

351 records

(Two xy locations
per record)

Level 1

One subroutine land

1 x 14 lands for subroutine

1 x 50 patterns

33 records

(Two xy coordinates
per record)

Level 2

One land subroutine

1 x 14 lands - pattern

1 x 5 patterns - set

1 x 1- sets - board

16 records

Level 3

Figure 7.7   Levels of IPC-D-350 implementation.

points correlate to flashes, and annotation records specify text that is restricted to the printable ASCII character set.

Since IPC-D-350 series has limited capability in defining lands, round, square, etc., positioned by a point record, this level allows for the use of subroutines to define nonstandard flashes (apertures) (Fig. 7.8). Nesting of subroutines is not allowed. The location description area, the "where-it-is" field (Fig. 7.9), is limited to two or three x,y fields; x,y,z fields are not supported until Level 3 implementation.

Level 2.   Level 2 implementation includes all of the requirements of Level 1 and also allows for the nesting of subroutines to one level. Subroutines can include information on special packages, patterns, paint-in routines, etc. (Fig. 7.10).

Round   —Feature—   Square   Multiple features defined as subroutine

Figure 7.8   Typical IPC-D-350 subroutine-defined features.

What it is	Detailed description		Where it is		User info
• Point • Lne • Annotation • Subroutine • Call subroutine	D-dimensioning or subroutine names L- layer S-node (electrical signal)	H-hole size or character height P-hole type G-generic (to futher describe points, line or annotation)	Three fields xy xy xy	Two fields xy xy	

Figure 7.9 Feature location records.

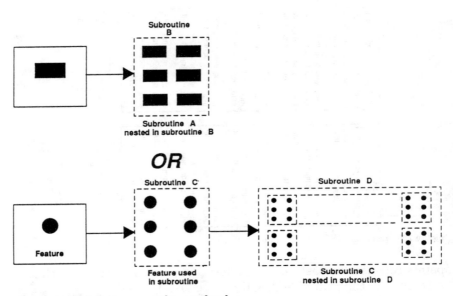

Figure 7.10 Subroutines nested to one level.

Level 3. This level is for the full implementation of the IPC-D-350 standard printed circuit board and artwork language. Level 3 allows for separate subroutines or for calling for separate external libraries. Subroutines can call one or more other subroutines that have been previously defined. Nesting of subroutines is permitted to any depth (Fig. 7.11). However, a subroutine cannot call itself or any other nested level that in turn calls a predecessor level.

7.8.1.3 Data sets

The data in the IPC-D-35X documents are formatted so that there is structure to the data that assist in the interpretation of the data for specific applications. The system also allows users readable comments to be interspersed within the files to facilitate human interaction.

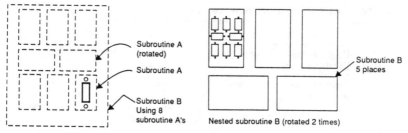

Figure 7.11   Subroutines nested to more than one level.

Subroutines are permitted to enhance the power of the data being communicated.

All of the data types start with a three-character code called the "op-code". For the 100 through 500 numbered data types, there are two sets of data for a complete record. The other sets of data are contained in what are called the "feature description area" (FDA) and the "location description area" (LDA). A feature location record specifies:

- A feature
- Parameters that describe that feature
- The location of the feature

Figure 7.9 illustrates the structure of feature location records.

The FDA is located in character positions 4–30; the LDA is located in character positions 31–71. Character positions 72–80 are for optional numbers or codes for use at the discretion of the CAD system, usually a record number. Table 7.4 shows the correlation between the record formats that are defined in each particular specification.

The electronic design description for a single design may contain different information at various points in the design cycle. Initially, printed circuit board description records may contain only board outline and blocked-area information, part description information in either library records or miscellaneous part drawing records, and electrical description records that describe the electrical relationships among the parts.

Once a design is completed, printed circuit board description records are supplemented with conductor routing information, hole information, and other data necessary to fabricate the board. The database can be added to as necessary in order to provide reference designator information for schematic drawing records, or any other data necessary for the intent of the user's database. Users of the IPC-D-350X series of specifications are encouraged to maintain data in a form that is

TABLE 7.4   IPC-D-350X Series Document Descriptions

Standard	Title Code	Content
IPC-D-350C	Printed Board Description in Digital Form	Artwork records and board description records
IPC-D-351	Printed Board Drawings in Digital Form	Schematic drawing, master drawing, assembly drawing, and miscellaneous part drawing records
IPC-D-352	Electronic Design Data for Printed Boards in Digital Form	Electronic description and bill of material records
IPC-D-353	Automatic Test Information Description in Digital Form	Electronic assembly testing format records
IPC-D-354	Library Format Description for Printed Boards in Digital Form	External library description and internal library description records
IPC-D-355	Printed Board Automated Assembly Description in Digital Form	Assembly data description records
IPC-D-356	Bare Board Electrical Test Information in Digital Form	Bare board testing (continuity, impedance, etc.)
IPC-DG-358	Guide for Digital Descriptions of Printed Board and Phototool Usage per IPC-D-350	User's Guide Handbook

self-sufficient and is not impacted by changes in supplementary data used in the design process. Thus, library description records may be repeated on archived data.

7.8.1.4   IPC-D-350[20]. IPC-D-350 is the most commonly used specification for transferring printed circuit board data (artwork, phototooling, and physical) from one computer environment to another. The data are in a "human-readable," 80-column card, ASCII format. Also, IPC-D-350 data can be prepared manually using a digitizer or from more sophisticated CAD equipment.

7.8.1.5   IPC-D-351[21]. The data contained in IPC-D-351 are intended to supplement IPC-D-350, specifically in the area of describing printed circuit board–related drawings in digital form. (Both of these specifications have been adopted for use by the Department of Defense.) The basic record formats in IPC-D-350 also apply to IPC-D-351. IPC-D-351 pertains to four basic types of drawings:

- Printed circuit board schematic diagrams.

- Printed circuit board master drawings.

- Printed circuit board assembly drawings.

- Miscellaneous part drawings.

7.8.1.6 IPC-D-352.[14] The information contained in IPC-D-352 is intended to describe the relationship between the elements used in the electromechanical design and packaging of electronic products using printed circuit boards as the major form of interconnection. Included in these descriptions are the logical and physical elements necessary as input to a design system, as well as the network or interconnection description between the various electronic parts. It is further intended that this structure provide the capability for describing all elements in their final form upon design completion. The concepts detailed in IPC-D-352 are supplemented by the descriptions defined in the other IPC-D-35X series documents.

7.8.1.7 IPC-D-353.[22] The information contained in IPC-D-353 is intended to describe the electrical and functional relationship between elements used in the design and packaging of electronic products. Included in these descriptions are the functional characteristics of both analog and digital components, their state and electrical requirements, as well as the network interrelationship descriptions between the various electrical elements that make up a functional electronic circuit or assembly.

The functional characteristics are described in digital form in order to enable data exchange between systems that support design, manufacture, assembly, and test. The information passed between these elements is only useful under the environmental conditions specified within the data. The format also provides a template for "self-test" generators. The following types of testing may be accomplished using the principles and formats defined in IPC-D-353 and companion IPC standards:

- Analog simulation

- Digital simulation

- Timing analysis

- Loading analysis

- Analog component testing

- Digital component testing

- In-circuit testing

- Bare-board optical testing

- Burn-in

In addition, IPC-D-350 may be used for

- Bare-board electrical testing
- Bare-board optical testing

7.8.1.8 IPC-D-354.[23]   The information contained in IPC-D-354 is intended to supplement the other standards in the IPC-D-35X series. It describes the use of libraries within the processing and generation of information files. The data contained in IPC-D-354 cover both the definition and usage of internal (i.e., exist within the information file) and external libraries. The libraries can be used to make data generated more compact and, thus, facilitate data exchange and archiving. The subroutines within a library can be used one or more times within any data information module. As a supplement, IPC-D-354 does not cover the specifics of the other standards in the IPC-D-35X series.

7.8.1.9 IPC-DG-358.[24]   The purpose of the IPC-DG-358 "User's Guide" is to enhance and elaborate the concepts detailed in IPC-D-350. It provides detailed step-by-step examples in all facets of using IPC-D-350 in order to define printed circuit board geometry and other vital characteristics. It also explains the flexibility of implementation and levels of IPC-D-350 to give the user a better understanding of the options provided.

IPC-DG-358 identifies those portions of IPC-D-350 that are machine-processable. Another purpose is to identify those portions that are intended to provide insight for the user who may receive a IPC-D-350 file and needs to produce a printed circuit board without any further instruction in paper form.

7.8.1.10 IPC-D-349.[25]   The information in IPC-NC-349 defines a machine-readable input format for computer numerical control (CNC) drilling and routing machine tools related to the printed circuit board industry. The format may be used directly to transfer drilling and routing information between printed circuit board designers, manufacturers, and users as the output CNC standard from converters that expand higher-level design input, such as IPC-D-350 data. It is intended to provide a common command structure that can assist in the printed circuit board manufacturing processes and is not intended to provide for every possible software enhancement.

7.8.2 IGES/PDES

The Initial Graphics Exchange Specification (IGES) is a communication file structure for data produced on and used by CAD and CAM systems in widespread use. It has been designed to serve as a receptacle for the data generated by commercially available interactive graphics design drafting systems. This structure provides a common basis for the automation interface.

IGES 1.0 was published as a National Bureau of Standards report in January 1980. It was recommended for standardization by the ANSI subcommittee Y14.26 in May 1980 and approved as ANSI Standard Y14.26M in September 1981. The most recent version of the standard, version 3.000, was published in April 1986. It has been approved as ANSI Standard Y14-26m-1987.

The complexity of the IGES project is enormous. This is due in part to the need for building in flexibility for future growth. To handle the complexity, future specifications will be developed using a three-layer methodology which divides the problem into three parts: logical, conceptual, and physical.

Information models are being developed for specific application areas, mechanical products, electrical products, architecture, engineering and construction, finite-element modeling, and drafting, and constituent technical areas, manufacturing technology, solid modeling, curve and surface modeling, and presentation data.

The methodology for data storage of IGES is an entity attribute database. Each record in the file associated with an entity contains parametric data related to that type of entity. Any entities that have a dependency relationship with other entities have within their records a pointer that defines that relationship. Extensions of IGES will include representation of the following design-related information:

- Integrated circuit design and connectivity
- Testing
- Simulation
- Inspection
- Hierarchical electronics system design.

The IGES organization is also involved with developing the Product Data Exchange Specification (PDES). This R&D effort addresses a different technology base than the IGES format. IGES is intended for information exchange between databases that must be interpreted by human beings. PDES is a complete database exchange that does not require human interpretation. The PDES initiation activities have been published and a strawman document will be ready by April 1987. PDES is the U.S. contribution to the international STEP (Standard for the Exchange of Product Data) development effort. The goal of the international effort is to develop a single standard for complete product data exchange.

7.8.3 VHDL[26]

VHDL, the "very high speed integrated circuit" (VHSIC) Hardware Description Language was approved as an IEEE standard (1076) in late 1987 and mandated by the Department of Defense under MIL-STD-454, Revision L, for application-specific ICs (ASICs) in early 1988. As such, it makes a significant impact on both IC and printed circuit board design.

The specific purposes of VHDL are:

- Provide a standard medium of communication for hardware design data.
- Represent information from diverse hardware application areas.
- Support the design and documentation of hardware.
- Support the entire hardware life cycle.

A fundamental tenet of VHDL is that the design specification should remain independent of its implementation. This pivotal rule focuses on the reusability of parts and ease of maintenance. The dedicated emphasis on commonality and reusability is structured to reduce the maintenance costs and effort demanded of manufacturers when parts are unavailable or when design revisions are needed.

VHDL is not the easiest design language to learn and apply to any specific function. Thus, there are instances in which another language may be more appropriate. However, for most designs, VHDL permits users to replace proprietary languages and apply top-down methodologies. Thus, it results in a system design that emphasizes behavioral aspects of a circuit or printed board.

The major portions of a VHDL description are entities, architectures, and configurations (Fig. 7.12). An entity is a design block or component and, at its lowest level, could be a single boolean gate. An entity declaration includes the number, type, and direction for each type of pin along with a series of parameters, called "generics," that are passed to the component model, called the "architecture." Entities can be thought of as schematic symbols; architecture, as a simulation model for a symbol; and configurations, as the link that associates the entity with one or perhaps several different architectures.

The packaging facilities inherent in the VHDL language, along with the features that permit abstraction of data, allow the user to unambiguously specify the way a design should function. Thus, the VHDL simulator provides very high quality feedback, optimizes design efficacy, and advances the usability of CAE tools for hardware designers.

```
USE std.std_logic.ALL;
ENTITY mc68020__ckt IS
--$ : VANTAGE__MFTACOMMENTS__ON
   PORT
           (
              PB0B,PB1B,PB2B,PB3B,PB4B,PB5B,PB6B,PB7B: INOUT t__wlogic;
              DS PORT A,DS__port__B,EXT__RESET: INOUT t__wlogic;
              DTACK__PORT__A,Dtack__port__b: INOUT t__wlogic;
              PA0B,PA1B,PA2B,PA3B,PA4B,PA5B,PA6B,PA7B: INOUT t__wlogic;
           );

END mc68020__ckt;

USt std.std_logic.ALL;
ARCHITECTURE mc68020 ckt OF mc68020__ckt IS
--$ : VANTAGE__METACOMMENTS__ON
-- Component declarations:
--
-- ( some component declarations removed )
--

   COMPONENT MC68020
      PORT
              (
                IPL0,IPL1,IPL2.AVEC,BR,BGACK,DSACK0,DSACK1,
                   BERR,CLK, CDIS: IN t__wlogic;
                IPEND,BG,SIZ0,SIX1,FC0,FC1,FC2,ECS,OCS,RM,DS: OUt t__wlogic;
                HALT,RESET,RMCDBEN,AS: INOUT t__wlogic;
                D31,D30,D29,D28,D27,D26,D25,D24,D23,D22,D21,D20,D19,D18,D17,D16,
                D15,D14,D13,D12,D11,D10,D9,D8,D7,D6,D5,D4,D3,D2,D1,
                   D0: INOUT t__wlogic;
                A31,A30,A29,A28,A27,A26,A25,A24,A23,A22,A21,A20,A19,A18,A17,A16,
                A15,A14,A13,A12,A11,A9,A8,A7,A6,A5,A4,A3,A2,A1,A0: OUT T__wlogic;
              );

   END COMPONENT;
   SIGNAL
           SGNL001011,DS,reset__pulse,sys,reset,SGNL001005,SGNL001002,
              data__dir,
           cpu__rw,ram__128k,SGNL000980,ram__96k,SGNL000981,
           ram__64k,SGNL000982,
           ram__32k,SGNL000986,oe,VUg2000__Addrl2J,
           VUg2000__Addrl3J,VUg2000__Addrl4J,
              -- ( some signal declarations removed )
BEGIN
-- Component Instantiations
-- .
-- . ( some component instantiations removed )
-- .
   mc68020__1: MC68020
      PORT MAP (
         A0 -> a00, A1 -> a01, A02 -> a02, A3 -> a03, A4 -> a04, A5 -> a05, A6 -> a06, A7 -> a07, A8 ->
      a08, A9 -> a09, A10 -> a10, A11 -> a11, A12 -> a12, A13 -> a13, A14 -> a14, A15 -> a15, A16 -> a16, A17 ->
      a17, A18 -> a18, A19 -> a19, A20 -> a20, A21 -> a21, A22 -> a22, A23 -> a23, A24 -> a24, A25 -> a25, A26
      -> a26, A27 -> a27, A28 -> a28, A29 -> a29, A30 -> a30, A31 -> a31, D0 -> d00, D1 -> d01, D2 -> d02, D3
      -> d03, D4 -> d04, D5 -> d05, D6 -> d06, D7 -> d07, D8 -> d08, D9 -> d09, D10 -> d10, D11 -> d11, D12 ->
      d12, D13 -> d13, D14 -> d14, D15 -> d15, D16 -> d16, D17 -> d17, D18 -> d18, D19 -> d19, D20 -> d20, D21 ->
      d21, D22 -> d22, D23 -> d23, D24 -> d24, D25 -> d25, D26 -> d26, D27 -> d27, D28 -> d28, D29 -> d29, D30
      -> d30, D31 -> d31, CDIS -> SGNL00594, DS -> ds, AS -> Address__strobe, DBEN -> dben, RW -> cpu__rw,
      RMC -> rmc, OCS -> ocs, ECS -> ecs, SIZ1 -> siz1, SIZ0 -> siz0, FC2 -> fc2, FC1 -> fc1, FC0 -> fc0, CLK ->
      Sys __ clock, RESET -> sys__reset, HALT -> pwr, BERR -> Berr, DSACK 1 -> dsack 1, DSACK0 -> dsack0, BGACK
      -> pwr, BG -> bg, BR -> pwr, AVEC -> pwr, IPEND -> ipend, IPL2 -> PIRQ, IPL1 -> pwr, IPL0 -> pwr );

CONFIGURATION VU8000 svgEnc68020__ckt OF mc 68020__ ckt IS
-- $ : VANTAGE__METACOMMETNS__ON
-- $ VU80000__sveErmc68020__ckt KNOWNAS
-- $ svg$mc68020__ckt;

FOR mc68020__ckt
-- Component specifications;
-- .
-- . ( some component specifications removed )
-- .
-- .
      FOR mc68020__1:MC68020
         USE CONFIGURATION work.mc68020__cfg
      GENERIC MAP (
         TIMING -> ''MC68020RC12'' ) ;
      END FOR:

END FOR:

END VU80000__svgErmc68020__ckt;
```

Figure 7.12 Example of VHDL entity-architecture configuration statement sequence.

7.8.4 EDIF[27]

The Electronic Design Interchange Format (EDIF) facilitates the movement of electronic design data between sophisticated databases in a commercial environment. This provides links between disparate systems as well as corporate control over information content. EDIF is intended for the representation of IC and electronic design data, not mechanical design data.

EDIF is capable of representing the electrical characteristics necessary for implementation of electronic products. It can express library/cell organization; provide extensive data/version control; describe the cell interface, cell details (contents), and technologies; and represent timing, geometry, and physical objects. Changes for version 2.00 include a number of extensions, changes and refinements to the format.

EDIF addresses the IC description necessary for fabrication, including modeling and behavioral aspects and some printed board descriptive material.

7.8.4.1 EDIF net lists.

A net list generally describes ports or pins of parts and their interconnections. EDIF net lists may also contain system- or user-defined properties. The properties may indicate electrical parameters such as timing, power rating, tolerance, capacitance, and loading. Net lists are commonly used to link a logical design system to a physical layout system.

Hierarchy. Net lists may be hierarchical or "flat." A complex hierarchical design is described as a set of interconnections between simpler subdesigns. Most schematic capture systems support hierarchical design, but most printed circuit board layout systems expect a flat description. Usually the schematic is flattened before being translated to EDIF format. Often this step is combined with the assignment of logical gates to physical packages.

Gates and swapping. Different CAE systems describe the same part in different ways. For example, the logic of a buffer device could be described as eight 3-pin gates, or as two 9-pin gates, or as one 18-pin gate. The choice of representation may limit the type of gate swapping that may be done during layout.

Data transfer is generally more successful if the circuit is described at the package level rather than at the gate level. Each system may then define its own mapping between gates and packages.

Power and ground. Some systems do not allow you to explicitly specify the connection of power and ground pins for ICs. The connections are made automatically, based on assumptions and naming conventions. Other systems may have different assumptions or may require explicit declaration of power and ground signals. The EDIF file should never rely on naming conventions. All assumptions should be eliminated in favor of explicit data.

Component databases. EDIF does not provide any standard part libraries. Different CAE systems use the different naming conventions to describe parts. Companies invest heavily in developing part libraries and may not be willing to abandon an existing library to accommodate a new EDIF link. Often the best solution is to use an auxiliary file to map parts from one database to parts in another database.

Electrical parameters. Different CAE systems model the same components with different types of electrical parameters. EDIF currently defines a very limited set of parameters and expected values. The EDIF specification allows user-defined extensions for this purpose. However, there must be prior agreement between the EDIF writer and reader to allow the user-defined properties to be correctly interpreted.

Engineering change. Design changes may originate either in the schematic capture system or in the layout system. These changes may include addition, deletion, or substitution of parts, changes to wiring, or changes to electrical parameters. EDIF does not provide a delta format for specifying only the changes.

Engineering changes are especially difficult if the logical and physical systems use different models for describing library parts. Many EDIF links provide no support for automatic engineering changes. As a minimum, the CAE systems should provide a way of comparing two versions of a design and reporting the differences so that the changes can be made manually.

7.8.4.2 EDIF schematics. An EDIF schematic includes all of the information that is in the net list. In addition, it provides a graphical representation of that data. A schematic diagram is often part of the final documentation for an electronic system.

Back annotation. After the printed circuit board layout is complete, the schematic is updated with information about physical packaging, including reference designators and pin numbers. In EDIF this information can be represented in different ways for hierarchical and flat schematics. Default information about the placement of this information can be included in the definition of library components.

Connectors. Some schematic connectors correspond to physical components on the printed circuit board. Others are used to show logical interconnections between pages or between hierarchical blocks. Still other "connectors" indicate global signals that are most commonly used for specifying power and ground nets.

Buses and bundles (signals and ports). The EDIF specification is extremely complex in this area. Signals and ports can be grouped together with names specified either as numbered arrays (buses) or as ordered collections. EDIF graphics describe the routing for the entire collection and for each individual line separating from the collection. An additional graphical object, called a "bus splitter" or "bus ripper," can be placed at each point that an individual line separates from the group. A bus splitter is treated as a special type of component on the schematic, but it does not correspond to any physical object on the printed circuit board.

Line types and fill patterns. EDIF defines a line or fill pattern as an array of dots or pixels. Most CAE systems do not provide that level of flexibility. Furthermore, the definition of a pixel is not standard across systems. Pixels are only useful for raster devices, not for pen plotters.

Grids. CAE systems use grids for defining the minimum spacing between design objects, points, and lines. The EDIF specification does not define this type of grid.

7.8.4.3 EDIF printed circuit boards. The EDIF printed circuit layout view is intended to include all of the physical design information for a circuit board, including traces, lands, targets, holes, planes, and vias. In addition, it incorporates all of the part and connectivity data that can be specified in the netlist view.

Holes. The EDIF standard does not include a mechanism for describing holes.

Hierarchy. The EDIF standard allows unlimited flexibility in the use of hierarchy and other grouping mechanisms for designs and subdesigns. Some of the hierarchical levels may correspond directly to physical objects such as components, pins, or lands. Other levels may exist only for convenience or to compactly represent data that are shared between different objects. Standard EDIF makes no distinction between convenience groupings and "real" design objects.

Layer types. EDIF graphical figures are grouped in logical layers. Each group can have its own color, line style, fill pattern, and text height. Several figure groups can be combined to produce one physical layer in the manufacturing process. However, the EDIF specification does not distinguish between the various types of layers needed in printed circuit board design and fabrication.

Negative layers. Some layers of a printed circuit board are typically defined negatively. For example, a power plane is usually described as a group of figures that describe the areas where there is no copper. The EDIF standard allows complex geometric combinations of figure groups to describe these layers: union, intersection, difference, and inverse.

Arbitrary rotation. Components mounted on printed circuit boards can be rotated any number of degrees relative to the board outline. The EDIF standard only allows 90° rotation angles and mirroring. If a component is to be placed at several different angles, the EDIF file must contain multiple definitions of the same part, differing only in orientation.

7.8.5 Computer-aided acquisition and logistical support (CALS)

The CALS program is a Department of Defense initiative to enable and accelerate the use and integration of digital technical information for weapon system acquisition, design, manufacture, and support. The benefits expected to be achieved by CALS include:

- Reduced acquisition and support costs
- Elimination of duplicative, manual, and error-prone processes
- Improved reliability and maintainability of designs directly coupled CAD/CAE processes and databases
- Improved responsiveness of the industrial base, i.e., to be able to rapidly increase production rates or sources of hardware based on digital product descriptions

As a result of CALS initiatives, the various design languages have been rated according to their capability for handling electrical product data descriptions and recommendations have been made as to how to implement and use these languages (Table 7.5).

TABLE 7.5 Design Language Process Functional Matrix

	System				Box				Board				Component			
	IGES	EDIF	VHDL	IPC	IGES	EDIF	VHDL	IPC	IGES	EDIF	VHDL	IPC	IGES	EDIF	VHDL	IPC
Behavioral Description																
1. General		●				●		●	●			●	●			●
2. Quality Level		●				●		●	●			●	●			
3. Signal		●	●			●		●		●	●			●	●	
4. Ports		●	●			●		●		●	●			●	●	
5. Quantitative performance		●	●			●		●		●	●			●	●	
6. Operating range		●	●			●		●		●	●			●	●	
7. Safety		●	●			●		●			●	●	●		●	
8. Simulation—behavioral		●	●			●		●			●			●	●	
Functional Description																
9. Functional partitioning	●	●			●	●			●	●			●	●		
10. Form factor					●				●				●			
11. Algorithmic description	●		●		●	●				●				●		
12. Interface control & limit	●			●	●		●					●			●	
13. Environmental test parameters																
14. Simulation & functional			●			●		●		●					●	
Logical Description (Digital)																
15. Symbol definition	●	●		●	●	●		●	●	●		●	●	●		●
16. Signal	●	●			●	●			●	●			●	●		
17. Ports	●	●			●	●			●	●			●	●		
18. Timing (description)	●	●			●	●			●	●			●	●		
19. Simulation—logic (including models)	●	●			●	●			●	●			●	●		

Circuit Definition (Analog)

20. Symbols
21. Gain charts/V-I plots
22. Frequency plots
23. Propagation delays
24. Timing description
25. Quantitative performance
26. Operating range
27. Q, R, & M calculations
28. Simulation (circuit)

Simulation

29. Fault Simulation
30. Test Vectors
31. Thermal
32. Vibration

Net List

33. Design Rules
34. Parts
35. Interconnectivity

Physical Design Layout

36. Physical design rules
37. Dimensions/tolerances
38. Package interfaces
39. Material properties
40. Reference designators
41. Cabling conductors
42. Detailed thermal analysis
43. Detailed R&M analysis

TABLE 7.5 Design Language Process Functional Matrix (Continued)

	System				Box				Board				Component			
	IGES	EDIF	VHDL	IPC	IGES	EDIF	VHDL	IPC	IGES	EDIF	VHDL	IPC	IGES	EDIF	VHDL	IPC
Physical Documentation																
44. Detail/package drawings	●	●	●		●	●	●			●		●	●	●		
45. Reference designators																
46. Dimensions and tolerances									●						●	
47. Material construction	●	●			●	●				●		●	●			
48. Assembly drawing and notes																
49. Parts list	●				●							●	●	●		
50. Fixturing	●				●										●	
51. Pattern geometry	●				●				●			●			●	●
52. NC data	●				●				●							
Assembly & Test																
53. Assembly specification	●	●			●	●				●	●		●	●		
54. Test/burn-in requirements						●					●			●		
55. Other Q, R, & M testing														●		
Installation																
56. Drawings				●				●				●				●
57. Tech manuals				●				●				●				●
58. Shipping container drawings				●				●				●				●

References

1. "The Most Commonly Asked Questions about PCB CAD," 1989, Personal CAD Systems, Inc., San Jose, Calif.
2. "Automated Design Guidelines," IPC-D-390, Revision A, February 1988, Institute for Interconnecting and Packaging Electronic Circuits, Lincolnwood, Ill.
3. William B. Holcomb, Singer Librascope, "Control of the Printed Circuit Through Design, CADS and Manufacture," IPC-TP-460, April 1983, Institute for Interconnecting and Packaging Electronic Circuits, Lincolnwood, Ill.
4. "Suggested Guidelines for Modification, Rework and Repair of Printed Boards and Assemblies," ANSI/IPC-R-700, Revision C, January 1988, Institute for Interconnecting and Packaging Electronic Circuits, Lincolnwood, Ill.
5. "Design Standard for Rigid Single- and Double-Sided Printed Boards," ANSI/IPC-D-319, January 1987, Institute for Interconnecting and Packaging Electronic Circuits, Lincolnwood, Ill.
6. "Design Standard for Rigid Multilayer Printed Boards," ANSI/IPC-D-949, January 1987, Institute for Interconnecting and Packaging Electronic Circuits, Lincolnwood, Ill.
7. "Printed Circuit Board Design—Illustrated User's Guide," 1989, Personal CAD Systems, Inc. San Jose, Calif.
8. "Design Standard for Rigid Printed Boards and Rigid Printed Board Assemblies," IPC-D-275, Spetember 1990, Institute for Interconnecting and Packaging Electronic Circuits, Lincolnwood, Ill.
9. "Guidelines for Phototool Generation and Measurement Techniques," ANSI/IPC-D-310, Revision B, December 1987, Institute for Interconnecting and Packaging Electronic Circuits, Lincolnwood, Ill.
10. "Printed Board Component Mounting", ANSI/IPC-CM-770, Revision C, January 1987, Institute for Interconnecting and Packaging Electronic Circuits, Lincolnwood, Ill.
11. "Component Packaging and Interconnecting with Emphasis on Surface Mounting", ANSI/IPC-SM-780, March 1988, Institute for Interconnecting and Packaging Electronic Circuits, Lincolnwood, Ill.
12. "Surface Mount Land Patterns (Configurations and Design Rules)", ANSI/IPC-SM-782, March 1987, Institute for Interconnecting and Packaging Electronic Circuits, Lincolnwood, Ill.
13. "Qualification and Performance of Permanent Polymer Coating (Solder Mask) for Printed Boards," ANSI/IPC-SM-840, Revision B, May 1988, Institute for Interconnecting and Packaging Electronic Circuits, Lincolnwood, Ill.
14. "Electronic Design Data Description for Printed Boards in Digital Form," IPC-D-352, August 1985, Institute for Interconnecting and Packaging Electronic Circuits, Lincolnwood, Ill.
15. "Checklist for Comparing CAD Systems and Software," Aptos Systems, *Printed Circuit Design*, January 1987.
16. "Guide for Digital Descriptions of Printed Board and Phototool Usage per IPC-D-350", IPC-D-358, Proposal, December 1989, Institute for Interconnecting and Packaging Electronic Circuits, Lincolnwood, Ill.
17. William Lange, Lange Associates, "D-350 Data Types and Structure," *Printed Circuit Design*, October 1986.
18. William Lange, Lange Associates, "Further Discussion of the D-35X Specifications," *Printed Circuit Design*, December 1986.
19. Dieter Bergman, Institute for Interconnecting and Packaging Electronic Circuits, "IPC-D-350: One Standard for PC Design", *Printed Circuit Design*, December 1987.
20. "Printed Board Description in Digital Form," ANSI/IPC-D-350, Revision C, March 1989, Institute for Interconnecting and Packaging Electronic Circuits, Lincolnwood, Ill.
21. "Printed Board Drawings in Digital Form," ANSI/IPC-D-351, August 1985, Institute for Interconnecting and Packaging Electronic Circuits, Lincolnwood, Ill.

22. "Automatic Test Information Description in Digital Form," IPC-D-353, April 1989, Institute for Interconnecting and Packaging Electronic Circuits, Lincolnwood, Ill.
23. "Library Format Description for Printed Boards in Digital Form," ANSI/IPC-D-354, February 1987.
24. "Guide for Use of Digital Descriptions of Printed Boards 'IPC-D-350' User's Guide," IPC-DG-358, May 1990, Institute for Interconnecting and Packaging Electronic Circuits, Lincolnwood, Ill.
25. "Computer Numerical Control Formatting for Drillers and Routers," IPC-NC-349, August 1989, Institute for Interconnecting and Packaging Electronic Circuits, Lincolnwood, Ill.
26. John Wiley and Gary Gordon, Vantage Analysis Systems, "VHDL Language Applications in Circuit and Board Designs", *Circuit Design*, November 1989, pp. 40–43.
27. Harvey Clawson, Hewlett-Packard Electronic Design Division, "Using EDIF for PCB Design," *Printed Circuit Design*, December 1987.

Layout and
Component Placement

8.1 Introduction[1,2]

The layout of the printed circuit board requires its own planning and design. Although CAD can automate printed circuit board layout more than it can automate logic design, most printed circuit boards require the judgment of the designer before layout with CAD begins. For instance,

- How many layers will the design take?

- Are there critical signals and parts that require special consideration?

- Are there peculiarities of the board that warrant special treatment?

- Where will potential problems come from?

- Which manufacturing process will be used to produce the boards after layout?

Most sophisticated electronic CAD systems provide the user with a combination of automatic and interactive design tools. The automatic tools allow the designer to accomplish certain tasks in a matter of hours that formerly took days or weeks. The interactive tools allow the designer to both fine-tune the results of the automatic tools and to accommodate in an optimal manner any technology details that are not handled automatically.

In the layout process, two essential tools are placement and routing. Placement positions a set of circuit elements, usually generated by packaging the output of a CAE schematic capture system onto the layout surface. Routing connects all electrically equivalent points in a legal manner.

Many users of CAD systems treat the placement and routing tools as "black boxes," which somehow mysteriously produce their intended output. If the system has good default parameters, these usually produce reasonable results. However, if the user has a good understanding of the behavior and characteristics of these tools, then the results can improve dramatically. By setting up the design to optimize the strengths of the tools and minimize the weaknesses, the user can work with, rather than against, the software.

8.2 Initial Considerations

8.2.1 Database preparation

Layout starts with a blank board outline created to your needs and specifications. The number of layers in the layout and the form factors of the board are specified in this initial database. The designer can usually create a library of blank board outlines, set up various layout design rules beforehand, and use them at layout time. This blank board is used with the schematic net list to build the starting printed circuit board layout database, which will then be analyzed, placed, routed, edited, and documented to produce the finished layout.

8.2.2 Layers

Layer-based CAD systems generally provide up to 100 layers that can be used for signal-carrying traces as well as for silkscreening, power and ground planes, and so on. The layers are usually referenced by name and each layer can often be assigned a different color for ease of use. For designs with surface-mount technology (SMT) devices, some CAD systems allow for layers on the top side of the boards to be mapped to corresponding layers on the bottom side, so that SMT parts can be freely moved from side to side during placement.

The essence of integrated CAE/CAD is, of course, that the schematic design can be directly used for printed circuit board layout without having to reenter the data for layout. The input to printed circuit board layout, therefore, is the schematic design net list. The schematic net list is first prepared for layout by packaging the logical gates into physical packages. Several circuit package programs automatically assign package reference designators to all unassigned logical gates in the net list and packages them into physical parts. However, they usually do not override any previously user-assigned reference designators in the schematic.

8.2.3 Symbol-to-part cross-referencing

CAD systems generally use some form of a symbol-to-part cross-reference table, supplied with the CAD library, to select the land patterns for each logical symbol. This makes it possible to select different packages for the same logical symbol, for example, SMT. Edge-board connectors are also automatically treated as components and signal input/output connector symbols used in the schematic can be packaged into edge connectors on the board.

8.2.4 Nonhomogeneous gates

Nonhomogeneous gates and gates that share common package pins can be automatically packaged by some CAD systems.

8.3 General Component Placement Considerations[1-3]

There are three classes of computerized placement systems, categorized by the degree to which they are automated, namely:

- Manual/interactive
- Assisted
- Automated

Although automatic placement and routing programs produce results rather quickly compared to manual methods, these results are often not as good as hand layouts. Since the computer is so fast, why not just try all combinations and pick the best one? Unfortunately, placement and routing belong to a class of problems that mathematicians call "np" (nondeterministic polynomial) hard, and it has been demonstrated that even fairly small np hard problems take an impossibly long time to solve (even using a computer) with an exhaustive search.

Many studies have been done to decide which type of initial placement is better. The consensus is that for large problems, spending the time to do constructive initial placement is worthwhile, while for smaller problems it is more efficient to choose the best of a number of random initial placements before going to the improvement pass.[4] All algorithms improve the initial placement by moving the components into "better" positions until the "best" placement results.

8.4 The Placement Problem[5]

The placement of components on a printed circuit board is crucial, but not as well defined. The layout designer must address complex considerations and constraints from a variety of concerns such as electrical

TABLE 8.1   Printed Circuit Board Design Problem Characteristics

Design problem	Solution
Analog circuits	
Off-grid pins	Ability to route to off-grid pins, gridless routing
No sharp corners	45° or any angle routing—manual or postprocess automatic
Shielding	Ground plane generation—pattern capability
Variable trace widths	Interactive trace width modification
High-speed circuits/ECL	
Stub length	Control/prevention in autorouter, reporting; disable copper sharing except where stub allowed
Parallelism	Control/prevention in autorouter, reporting
Source & termination spec.	Specification in schematic or netlist; autoroute in daisy-chain order
Shielding	Ground plane generation
Surface mount design	
Components on both sides	Special features for placement & routing
No through holes assumed	Router must add vias
Component footprints	Component creation features and copper sharing in the router
Buried vias	Special router features
Multilayer	True multilayer router
Multilayer boards	True multilayer router
Large boards	
Capacity	Memory space
Interactive performance	Graphics acceleration
Autorouting performance	Routing acceleration, rip-up and reroute
Military/aerospace	
No vias under components	Via exclude for components
No vias at all	Router controls
No pads on innerlayers	Post process deletion of pads, autorouter recognition of available space
Buried vias	Special router features
Odd board geometries	Drafting facilities
Any angle components	Any angle placement
Manufacturing technology	
Fine lines	Gridless routing, fractional grids
Irregular/mixed grids	Gridless routing, multiple grids
Connectors/pin grid arrays	Gridless routing, ability to route to off-grid pins
Designer efficiency	Rip-up and reroute technology

characteristics, manufacturability, testability, reliability, maintainability, and even aesthetics (Table 8.1).

Although each area of concern is valid and important, they often conflict, leaving the layout designer with no clear direction. In addition, it is often difficult for the layout designer to determine the effectiveness of a given placement because certain variables cannot easily be discerned at the placement stage. These include routability and internal packaging (gate and pin assignments).

The task of printed circuit board placement spans the fields of CAE, CAD, and CAM (Fig. 8.1).[1] To place a board well, one needs to understand the peculiarities of the given design, the strengths and weaknesses of the router to be used, and the manufacturing process. Because circuit design, layout, and manufacturing of printed circuit boards are most often performed by separate departments, such a person is hard to find. Instead, there is a communication problem that focuses on the placement.[6]

It is high priority for the layout designer to adhere to the circuit designers' concerns during placement. Typical circuit design concerns are that critical signals must be short, ECL terminating resistors must be placed at the end of the daisy chain, bypass capacitors must be interspersed evenly between the DIPs, and thermally critical components must be placed at specific locations. Some other complications are

- All devices must be within a certain area.

- Some devices are prepositioned and cannot be moved; these are commonly called "fixed components."

- Some devices can only go in certain areas of the board surface.

- Some or all devices may have specific locations that are legal, for example, in a gate array.

- Some devices need to be placed near other devices, such as filter capacitors and resistors, which need to be near DIPs.

- Some devices can be repackaged, the gates within components or macros can be reassigned.

- Some connections, such as clock signals or signals in tight timing paths, must be made as short as possible.

- Some devices require that there be a certain clearance around them.

These complications only limit the possible solutions. The exact mechanisms for accommodating these complications vary. The basic problem is to find the best possible placement amongst the legitimate possibilities. Since it is known to be impossible to find the best placement for any large number of devices, all automatic placement routines attempt to derive and/or refine placements heuristically. This is most often done by starting with a placement and continually trying to find a better placement.

If all of the above is not enough, reliability and maintainability issues impose additional demands. When adequate cooling is a consid-

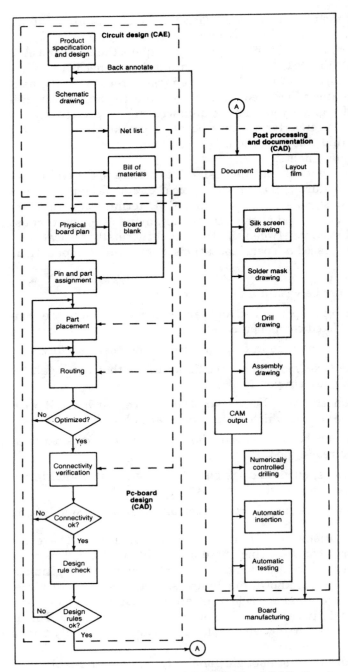

Figure 8.1   CAD cycle.   *(Courtesy of Electronic Trend Publications.)*

eration, as often is the case with today's increasing densities, component location and orientation should provide for air vents or channels to accommodate the flow of air and heat sinks. Thus, thermal analysis programs are sometimes used to measure temperatures after placement. Increasing reliability concerns indicate that thermal considerations should be integral to the placement process. From a maintainability viewpoint, if a component fails, it should be easily accessible for test probes, removal, and replacement.

To get a feel for the time involved, consider a small placement problem involving 100 circuit elements and 100 sites. The number of combinations can be computed as 100 choices for the first site, multiplied by 99 choices for the second site, multiplied by 98 choices for the third site, and so on. This number, called 100 factorial, has the approximate value of 10 raised to the 158 power, that is, 1 followed by 158 zeroes.

In comparison, the fastest supercomputers execute several billion instructions per second which, if they ran continuously, would result in the execution of about 10^{16} instructions in a year. Clearly, this number is miniscule when compared to 100 factorial!

8.4.1 Analog circuits

Analog circuit designs may utilize components with irregular shapes and dimensions in addition to conventional IC components. Also, analog circuit component leads will sometimes be off-grid and not easy to route automatically without the use of off-grid or gridless techniques. Thus, the ability to use interactive aids is important.

The use of different conductor widths is common and the interactive tools should allow the widths to be easily changed. It is also helpful to provide "true" size and shape display for accurate visualization of the layout. The sharpness of a turn in the conductor may be a critical analog circuit design parameter, and curved conductors or various angles might be required. Another analog circuit consideration is the common need for special shielding provisions.

8.4.2 High-speed circuits

High-speed circuits, such as those using emitter-coupled logic (ECL), advanced-Schottky transistor-transitor logic (TTL), or gallium arsenide (GaAs), often require special printed circuit board design considerations. Their operational frequencies may cause them to act as transmitters or be susceptible to radiofrequency interference (RFI). This elevates the importance of component placement and, as with analog circuits, introduces the need for shielding devices.

8.4.3 Surface mounting

SMT printed circuit boards best illustrate the challenges of the placement process. SMT components come in a variety of sizes and shapes, and may be placed on both the top and bottom of the board. Conventional through-hole components are often used on the same design. Just fitting the components on the board and ensuring that there is no package or pin interference can be a problem.

8.4.4 Large boards

Large printed circuit boards may also pose special problems for a design system. Limitations on the number of components, connections, or board dimensions may require splitting the board in half or thirds for individual design with the artwork combined later. This technique makes it difficult to optimize component placement.

8.4.5 Manufacturing

Manufacturing and test processes dictate other placement criteria. The placement should support and optimize automatic assembly, particularly if surface-mount devices are used. There are many constraints and considerations, depending upon the particular machines and processes, such as space between components for grippers and special handling for certain orientations. Alignment of the components is also an important consideration for wave-soldering. If automatic testing is planned, all component pins should reside on a common pin grid—typically 100 mils. (This is one constraint which complements routability.)

8.5 Metrics[2,6,7]

One of the difficulties in finding a "better" placement is that the only true measure of a good placement is its routability. The only way to measure routability is to route the design. It is much too expensive to find a good placement by routing each of many possible alternatives. Therefore, a number of attempts have been made to develop criteria by which the placement can be evaluated quickly.

The designer can qualitatively assess routability by analyzing a "rat's-nest" display of all connections (Fig. 8.2).[4] By sorting the components to reduce crossovers and length of connections, designers can follow their own intuitive cost functions. A cost algorithm can evaluate placement and interconnection problems, thus providing a mechanism for comparing the cost effectiveness of one placement to an-

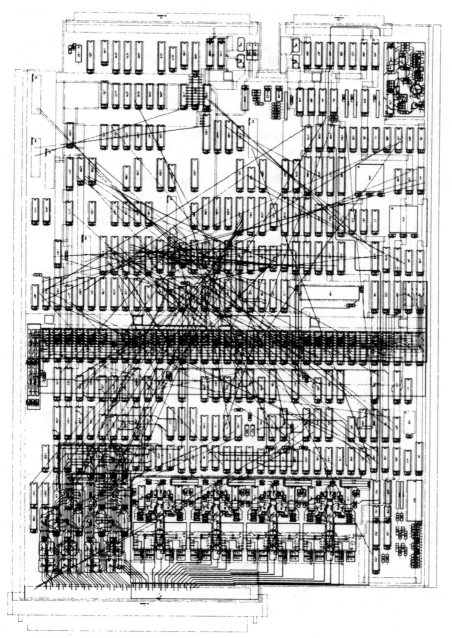

Figure 8.2 "Rat's nest" wiring. *(Courtesy of Computervision.)*

other. The algorithm uses net lengths, crossovers, available wiring (density), etc., to make this determination.

The other various methods of quantitatively evaluating a placement are called "metrics." Commonly used metrics are wire length estimation, min-cut, and node congestion.

8.5.1   Wire length

Routability is often measured as a function of total wire length and an estimate of the number of vias required. In addition to the visual feedback from rat's nests, route densities can also be graphically displayed as histograms, along with a figure of merit that indicates relative routability improvement. As parts are moved, or parts, gates, or pins are swapped, the histogram is updated to reflect whether the new placement is better or worse, and by how much. Any of the swaps can often be easily reversed with a command that restores the condition prior to swapping. The other methods of estimating the wire length for a net are called "half-perimeter," "modified-Steiner," and "min-span."

8.5.1.1   Half-perimeter.   The half-perimeter method draws the bounding box around the pins in a net and adds the width and height of the box.

8.5.1.2   Modified-Steiner.   The modified-Steiner method draws a horizontal line at the average height of all the pins from the left side of the box to the right side of the box using the bounding box again (assuming the box is wider than high). Now, the modified-Steiner distance is the length of the horizontal line plus the length of each vertical segment required to connect each pin to the horizontal line.

8.5.1.3   Min-span.   The min-span method reduces the net to pin pairs using the treeing technique described in the section titled "Network Treeing." The sum of the Manhattan distances (half-perimeter of pin pair's bounding box) of all pin pairs is the min-span estimate of wiring length.

8.5.2   Min-cut

The use of min-cut algorithms is quite simple. They measure the number of nets that cross a defined set of cut lines. The cut lines are normally drawn between rows or columns of devices because the value of a metric depends on how closely it approximates the routability of the design.

First, the components are divided into two groups. Swaps occur between these groups, with the cost function being the connection between the groups. When minimized, each group is then broken into two more groups and the process continues. At some point, there is only one component per group and the analysis is completed.

These types of algorithms may be quite fast, but they tend to work poorly when there are fixed components on the board. The routability depends on how many nets need to be routed through the routing resources.

8.5.3 Node congestion

The "node congestion" metric analyzes the number of nodes (pins on components, ports on macros) within each region of the routing area. Node congestion attempts to equalize this number across all regions.

8.6 Manual/Interactive Component Placement

The manual/interactive approach provides designers with the ability to graphically manipulate the placement of the components on the board until they are satisfied with the layout. The quality of the placement in this approach depends entirely on the designer's experience, expertise, and judgment. In practice, some components will normally be positioned by automatic procedures and others by designer-defined position entry. That is, the designer can define positions, and later initiate computations so that the placement can be made with respect to the prepositioned components. Alternatively, automatic placement can be initiated for only some components, with the remaining components positioned by the designer at a later time.

The designer can, at any time, display all the connections of components already placed during the component placement phase. The resultant screen displays a rat's nest of straight lines drawn from pin to pin in all directions (see Fig. 8.2). The reason for displaying the rat's nest is to give the designer a picture of possible congested areas on the board that can be eliminated before automatic routing starts. Components can be manipulated singularly, in groups, or placed in rows or columns in order to meet design requirements or to ease the routing.

8.7 Assisted Placement[2]

The assisted approaches include those in which the system, to varying degrees, helps the designer make the placement. One of the simpler assisted techniques is the presentation of the point-to-point connection (rat's nest). One merit of the rat's-nest approach is that the designer

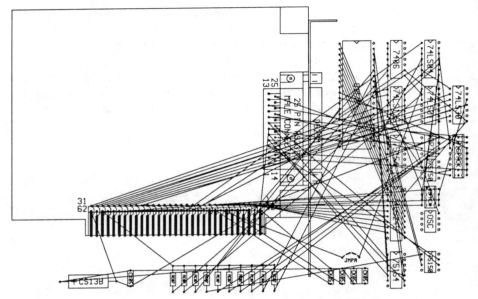

Figure 8.3 Preplacement and rat's nests. *(Courtesy of Personal CAD Systems.)*

can see simultaneously on the video screen all of the board connections to be made. However, systems handle the presentation of the rat's nest differently. Some systems can display a single net, thus suppressing the nets that the designer does not need to view, while in other systems, a net list manipulation is necessary in order to obtain the more selective approach.

One assisted method allows the designer to display some or all of the components at the edge of the work area during the placement process (Fig. 8.3).[1] The visual display of the components aids the component placement for the designer. In the assisted method, the designer allows the computer program to make an initial random or constructive placement just to get started.

The computer program may also be used to improve or refine a placement that the designer did manually or with system assistance. The designer then attempts to improve upon this placement by moving parts to more appropriate locations. Some systems assist the designers by using a histogram that identifies high-density conductor areas. Moving the components to different locations optimizes the wiring density.

8.8   Automated Placement[5]

The automatic approaches, as their names imply, automatically place the components of the circuit on the board. The designer must provide some placement parameters, and after the computer placement is complete, the designer is allowed to manually change the results. Most automatic placement routines (algorithms) automatically place the components at strictly defined locations on the printed circuit board; other systems allow random positioning of components, restricting only component body interference. All systems attempt to minimize the length of the connection paths. Some routines also consider special design rules for specific designs by measuring the total orthogonal distance (Manhattan distance) between all points of each net connection.

8.8.1 Basic Considerations

The relevant considerations for automatic placement tools fall into two categories: topological and geometric.

8.8.1.1 Topologic concerns. Topologic concerns are most commonly addressed in automatic placement. They are derived from the net list and deal with the position of each component realtive to the others. The goal is to optimize wire length and routing density.

8.8.1.2 Geometric concerns. Topologic considerations alone are not adequate to address the needs of placement. The more important considerations are geometric, i.e., those that are derived from the shape of the components and the board onto which they will be placed. Geometric considerations also include how the board space is used and where exactly on the board each component is positioned. The most critical geometric constraint is that all components must fit within the usable board area without violating the design rules. Other constraints are

- The components must be spread throughout the board area to prevent unnecessary routing congestion and to facilitate automatic assembly.

- If possible, the components should be aligned to create air vents, accommodate automatic test equipment, improve soldering performance, and provide a more aesthetic layout.

Changes in the printed circuit board industry demand a more serious approach to the subtle geometric aspects of component placements. These include

- Advances in routing and manufacturing technologies make very dense boards possible. To ensure that all components fit properly in a dense board without degrading the wire length, explicit algorithms are required.

- The advent of SMT designs introduces further complexities. Components can be placed on both sides of the board; they also come in a variety of shapes. Automatic assembly is mandatory for most SMT designs. Thus, manufacturability issues must be taken into account.

- In today's large designs, the fine-tuning of placement requires automatic assistance. In terms of time, manual fine-tuning is no longer affordable, as it was in smaller designs.

- Increasingly complex technologies, such as ECL, demand more designer involvement in the placement process. To save the designer's valuable time, more comprehensive placement tools are needed.

8.8.2 Heuristics

Automatic layout programs use "heuristic" and "local optimization" techniques to produce reasonable results in a reasonable amount of time. A heuristic can be thought of as a good rule of thumb. For example, a heuristic often used by automatic routers is to route shorter nets before longer nets. Local optimization can best be explained by using an example from placement. Assume the program is given some arbitrary placement; the program attempts to improve the placement by swapping circuit elements; however, it will only perform a swap if the resulting placement is better.

The advantage of this technique is that it improves computational speed by limiting the number of options that are examined. The disadvantage of this technique is that it produces a "locally optimized" rather than "globally optimized" layout. By changing the initial layout or by allowing some swaps that temporarily result in a worse layout, a better overall result might be obtained.

8.8.3 Placement strategy

The wide range of local concerns that a placement algorithm must address makes a general and comprehensive solution intractable. On the other hand, effective algorithms exist for several specific placement tasks. What is needed is a sensible and effective coupling of human intelligence and the computational power of a machine. The designer can bring into the process experience conventions and an understanding of the design's structure and data flow. The automatic tools should perform the computation-intensive optimizations, analyze the non-visual properties of the placement, and assist the designer in the routine manipulations of components. This approach is sometimes referred to as an "expert system."

In order for the designer to adopt the tool for everyday use, it must have the following characteristics:

- Be easy to use.

- Provide avenues of communication from the schematics to the placement tools by tagging the schematics with parameters. Some parameters reflect generic electrical corners, such as ECL termination, power, and signal priority. These issues should be treated explicitly by the placement algorithms. The circuit designer should also have the ability to specify arbitrary parameters that are used only as "hooks for manipulation" at the placement stage.

- Provide a set of automatic placement tools that perform powerful manipulation on portions of the design. These tools include functions that place a portion of the design in a designated area, swap gates and pins, spread and align a region of the board, and others. These tools must be flexible and reentrant, allowing the layout designer to easily coordinate their invocations.

- Provide a set of analysis tools that measure aspects of the placement quality and indicate trouble spots. They should include density display as a routability measure, manufacturability, and thermal analysis.

- The placement algorithm should incorporate not only generic geometric and topologic considerations but also certain specific electrical considerations. For example, the algorithm should delay the placement of the ECL terminating resistors until the end. Then, it should place them as close as possible to the input end of the daisy chain. ECL nets should be prioritized to reduce their length at the expense of non-ECL nets.

- Special effort must be taken to ensure that a minimum number of components are created in the course of the packaging, without compromising the wire length. The minimal packaging can be achieved by extensive bookkeeping, focusing on the number of computers created, their functional composition, and common-control pin utilization.

- If the design has power/ground signals assigned to a split plane, the algorith should place components over the appropriate plane areas.

- SMT component placement must be supported intelligently on both sides of the board, even on boards of mixed technology (boards with both SMT and through holes).

- Shape-to-shape clearances should be maintained to facilitate automatic assembly, testability, reliability, and maintainability.

8.8.4   Placement features[3,8]

8.8.4.1   Renaming.   In many systems, a component designator can be changed to a new name by entering a record at the keyboard, or on some systems from a menu at the graphics terminal. Components are renamed, usually left to right, top to bottom, ascending numerical sequence within a specific reference designator class to facilitate test and checkout. This resequencing of the reference designators along with any component, gate, or pin swapping can then be back annotated to the schematic.

8.8.4.2 Matrices[7]. In many CAD programs, components can be positioned by defining a matrix of columns and rows. Each column is defined by an X coordinate and each row by a Y coordinate. The use of these tools is slightly more complex than the others. However, by specifying row and column information, a designer can quickly construct a matrix of components.

Components placed with this type of feature generally need some optimization later, because they are positioned in no special order. One drawback to matrix placement tools is that many components do not fit well into a regular arrangement. Other methods used to position in columns are to define a matrix with step increments for the placement of the components, define columns and rows by specific irregular increments, or by system default columns and rows.

8.8.4.3 Function assignment. Some CAD systems allow you to define functions of components as fixed and free functions and then execute commands that automatically swap those free functions between component packages for the best possible placement and shortest connectivity length. In other systems, function assignment is done on the schematic or in the net list, and any swaps must be edited.

8.8.4.4 Swapping. Another possible feature of the CAD design systems is the ability to swap the positions of like components so that the possibility of 100 percent interconnection is enhanced. Some autoplacement package provides the capability to swap parts (Fig. 8.4), gates (Fig. 8.5), and pins (Fig. 8.6).[1] Components and pins can be swapped automatically in order to obtain the shortest possible overall wire length, and components, gates, and pins can be swapped interactively to minimize routing complexity and to shorten overall wire length.

In pairwise interchange, a component is swapped with each other component of its type or class, and a computation is performed to determine whether total connection length is reduced. If the connection length is reduced, the swap remains; otherwise the components are returned to their original positions. This exchange continues until all components have been tried.

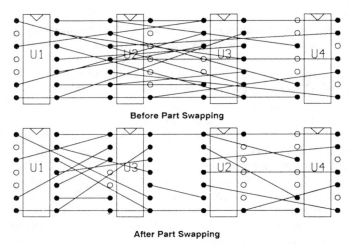

Before Part Swapping

After Part Swapping

Figure 8.4   Part swapping.   *(Courtesy of Personal CAD Systems.)*

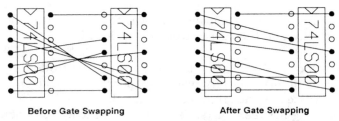

Before Gate Swapping   After Gate Swapping

Figure 8.5   Gate swapping.   *(Courtesy of Personal CAD Systems.)*

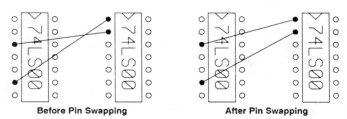

Before Pin Swapping   After Pin Swapping

Figure 8.6   Pin swapping.   *(Courtesy of Personal CAD Systems.)*

8.8.4.5 Moving and rotating.   This is usually an interactive design tool accessed from a menu during the design placement phase. It enables the designer to move and or rotate any individual part in the design. Smarter software packages also have the ability to move and rotate a complete group, or to rotate a complete component type.

8.8.4.6 Partitioning. In the board-partitioning placement technique, the design is reduced to a sequence of "partitioned" problems. First, a vertical boundary line splits the set of component locations into left and right halves. The set of components is then partitioned into halves; the partitioning procedure swaps component groups and components to minimize the crossing count for the vertical boundary line. Next, the set of locations is split into upper and lower halves separated by a horizontal boundary line, and the same component swapping method is used with an additional constraint: All left or right assignments previously developed must be respected (see Fig. 8.7).[8]

Each component has now been placed in either the left-upper, right-upper, left-lower, or right-lower quadrant and the placement of that component has been optimized with respect to the "crossing count" min/max criteria. This partitioning technique is then applied to each of the quadrants in turn, and then to the resulting octants until each region contains one board location.

8.8.4.7 Force vectors[2]. Another useful decision-making tool provided by some CAD software is the use of force vectors. These are lines drawn from the center of each component to a theoretically perfect placement location based on the current component locations.

8.8.4.8 Placement lattice[2]. Automatic placement can be controlled in many ways. Placement orientations are user-specifiable. You can specify placement lattices or grids that tell the autoplacer the horizontal and vertical spacing between parts. Several sets of lattices can be specified for placing different classes of ICs and types of parts, 14-pin DIPs, 40-pin DIPs, pin-grid arrays, etc.

8.8.4.9 Cutlines[2]. Some placement programs allow the designer to specify horizontal and vertical cutlines to partition the board. This tells the autoplacer to place heavily interconnected parts into localized areas indicated by the partitions.

8.8.4.10 Capacitor placement[7]. Another placement feature is the capacitor placer. This alogorithm allows the designer to specify a relative placement between ICs and capacitors. It then places the capacitors accordingly.

8.8.5 Iterative placement[7]

The most basic approach to automatic placement is to start with a placement, apply the metric of choice to obtain a "score" for the current placement, and then iterate the following operations:

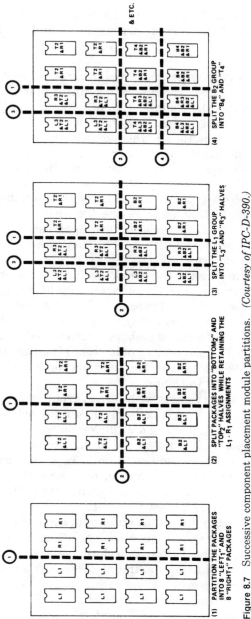

Figure 8.7 Successive component placement module partitions. *(Courtesy of IPC-D-390.)*

251

- Pick a new, legal arrangement of devices.

- Obtain a score for the new arrangement.

- If the new score is "better" than the current score, then make the new arrangement the current arrangement and the new score the current score.

When this approach is stopped, assuming some better placements were found, the designer should have a better placement. Even this simple approach has the complication of making sure that the same arrangement is never tried more than once.

A much more serious flaw is the slowness with which this system will arrive at an acceptable answer. It is slow because all of the choices for a new arrangement are random. Thus, some placements that a human would discard as obviously wrong, the computer program will score. Scoring takes time. Therefore, a number of heuristics have been developed that improve the likelihood of a new arrangement being better than the existing arrangement. This modification results in step 1 becoming pick a new, legal, likely to be better arrangement of the devices.

The problem with most of the heuristics is that they only change the design a little each iteration, most commonly by swapping two components or macros or gates. Therefore, if a sequence of moves would result in a better placement, but the first few moves of the sequence result in a worse placement, the above heuristic will not allow the set of moves to occur. This situation would occur if two groups of four components need to be exchanged but the four components within each group have many nets in common.

The problem is that the nets within each group will get longer if one component from each group is exchanged (or more nets will cut the cut lines), so that change is discarded as undesirable. This phenomenon is commonly referred to as "finding the local minimum." An analogous situation would be to find the lowest point in a square mile area and while only walking downhill, that is, only accepting better moves. Clearly, this might end up in a small indentation and not allow going up a few feet in order to go down 20 feet.

There are a few techniques that attempt to avoid the local minimum problem. Each of these techniques are variations of the simple technique stated above. The following techniques are iterative improvement techniques. The analogy of finding the lowest point will continue to be used to help you see how these techniques would be useful. However, all of these techniques involve working with many more placements than the above technique. Therefore, these techniques require more run-time for the program. Commonly the user is asked to give parameters to the programs to help make the trade-off between time and the quality of the solution. It should be noted that almost always there is a point of diminishing returns, that's the point where the algorithm should stop. The designer is also asked to give parameters to determine when the point of diminishing returns has been reached.

8.8.5.1 Scoring. One iterative improvement technique merely scores the placement every so many moves. Therefore, if the heuristic for choosing the moves is reasonably smart, multiple move sequences, such as the one mentioned above, are possible. In this analogy, the altitude would be noticed every, say, five steps.

8.8.5.2 Passes. Another iterative improvement technique works in passes. According to a preset heuristic, the program methodically works through an entire set of moves, scoring the placement after every move. At the end of the pass, the best placement of all moves will be accepted. If the beginning placement was the best, the program stops; otherwise, a new pass is initiated. The program continues until no progress is made, or a prescribed number of passes are executed.

In the analogy being used, it might be necessary to decide to go 1 step north, 2 steps east, 3 steps south, and 4 steps west, noticing the altitude after every step, or each sequence of steps chosen. At any rate, it can be seen that the algorithm is likely to get out of any little indentation. To get out of a small valley, it would have to increase the size of the spiral, and could well end up walking up a hill for a long time. This is the speed-quality trade-off; the longer the sequence of moves, the more time, but the better result.

8.8.5.3 Simulated annealing. A third iterative improvement is called "simulated annealing" because it was motivated by observing how metal anneals and what a smooth (optimal) solution is produced. The difference is in deciding whether or not to accept a new placement. There is always a value that injects a level of randomness into the process. When the level of randomness is low, very few worse placements are accepted. When the level of randomness is high, many worse placements are accepted, and only the much-worse placements are rejected. As a rule, the randomness is set high early in the process and is then increasingly lowered as the process continues. Step 3 in the iteration loop becomes: "If the score is better, accept the new placement; otherwise, depending on level of randomness and how much worse the placement is, accept or reject the placement."

In the analogy being used, it would wander aimlessly for awhile, and then start being more and more aware of the goal to try and find lower ground. In the end, it would only go downhill.

8.8.5.4   Metric deformation.   The last technique is to deform the metric every so often. For example, if a wire-length metric was being used, the y value could be multiplied by 1.25 and the x value by 0.75. Using this metric, several moves, that in reality are yielding "worse" placements, would be accepted because they look better under the deformed metric. In the analogy, it would be able to determine when altitude went awry every so often. It would then climb out of valleys and hopefully find deeper valleys.

Because the above techniques are all heuristic, some work better than others on certain types of problems. However, they all tend to get reasonable results that are better than manual techniques for very large problems because of the immense patience of the computer. They are not better than the manual techniques on medium-size problems because of the human problem solver's insight.

8.8.6   Constructive placement[7]

Another method is the "constructive placement" technique. This placement approach starts with a set of devices and derives the placement. A common technique is to iteratively partition the devices into smaller and smaller groups. The groups are associated with certain areas of the routing surface. Partitioning is a "min-cut" approach that attempts to minimize the number of nets that exist in more than one partition.

There are many techniques for doing this, but the basic idea is fairly intuitive. It simply divides a set of components into two groups, relative sizes of the groups determined by the corresponding relative sizes of the areas on the routing surface, such that the fewest number of nets exist in both groups. Then for each group, it partitions again. This process continues until the groups are reasonably small. Then the devices are placed on the legal locations within the corresponding area.

Any fixed component must always be in the group corresponding to the area the component is in. If there are only a certain number of legal locations for a certain device type and there are that number of devices of that type, the partitioner has set up the groups such that there are enough locations to accommodate the devices in each group.

Finally, it is possible to end up with a device that does not fit into its final group. In this case, the placer attempts to locate the device in the area corresponding to the previous group to which that device belonged in the process, and so forth until a home is found for the device. (Before the device was assigned to this group it belonged to a group that was split into two parts, its current group being one of the two.) If the device cannot be placed anywhere on the routing surface, it is considered to be "unplaced."

In the case of associated devices just mentioned, the process will generally only be applied to devices in dual-inline packages (DIPs). After the DIPs are placed, the associated discrete components will be placed with respect to the DIP locations. Most commonly, constructive placers are augmented with an iterative improvement placer in order to get the most routable design.

References

1. "Printed Circuit Board Design—Illustrated User's Guide," 1989, Personal CAD Systems, Inc., San Jose, Calif.
2. Robert Dean and Howard Schutzman, "A Layman's View of Placement and Routing Algorithms," Personal CAD Systems, Inc., San Jose, Calif.
3. "The Most Commonly Asked Questions about PCB CAD," 1989, Personal CAD Systems, Inc., San Jose, Calif.
4. Robert Huelsdonk, Honeywell Marine Systems Division, "Printed Wiring Board CAD," *Scientific Honeyweller*, vol. 7, no. 2, Summer 1986.
5. Gabriel Bracha, Jerry Harvel, and Assaf Dvir, Daisy Systems Corp., "The Pros and Cons of Automatic Component Placement," *Printed Circuit Design*, March 1988, pp. 17–20.
6. Bob Milne, "Routing Your Way Through PC-Board Design Tools," *Electronic Design*, vol. 36, no. 10, April 28, 1988.
7. Ben Meyers and Diane Elmer, Scientific Calculations Inc., "An Overview of Placement Algorithms," *Printed Circuit Design*, March 1988, pp. 13–16.
8. "Automated Design Guidelines," IPC-D-390, Revision A, February 1988, Institute for Interconnecting and Packaging Electronic Circuits, Evanston, Ill.

Conductor Routing

9.1 Introduction[1]

Manual routing of the interconnections on a printed circuit board uses a combination of tactics and strategies which are not easily formulated for execution on a computer. Flexibility in the approach and changes in tactics occur frequently as board design progresses. No single mathematical algorithm can hope to successfully emulate the problem-solving ability of an experienced professional. A good designer uses all known strategies in order to evaluate and solve each individual connection problem.

To manage this automatic routing, tools must meet three requirements: fast results (ideally to 100 percent completion), manufacturable board designs (e.g., low cost and high yields), and final products that conform to design restrictions. These requirements are not easy to meet, given increasing printed circuit board size and the variety of techniques that may be used to manufacture them.

To achieve the goals of reduced design time, improved manufacturability, and total conformance, designers need more than just technical innovation in an automatic router. Designers must examine its performance in relation to their specific design problems. How flexible is the router in meeting design rules? To what extent can routing be customized to conform to specific design and manufacturing requirements? Will it handle next month's, or next year's, designs as well as the boards currently being manufactured?

9.2 The Routing Problem[2]

Advanced printed wiring board technologies and their associated design requirements constantly are being explored and developed, and with them, new printed circuit board conductor routing technologies.

It is important to discern the differences between new algorithms and new implementations, and to compare the benefits of each in terms of speed, route quality, and applicability to a particular design problem.

A well-implemented routing database can keep track of all the components on the board, and the nets they belong to, thereby permitting support of features such as copper sharing and routing by net. These two features are designed to improve the quality of automatic routing through the use of additional information from the schematic or netlist. The routing problem can be divided into four subproblems:

- Network treeing
- Ordering
- Routing
- Post-route processing

9.2.1   Network treeing[3]

Many routers work from a single source to a single target. Therefore, a network with more than two pins is often subdivided into pairs of pins (Fig. 9.1). A common technique for pairing pins is known as "min-span." Basically, the algorithm generates pairs such that the sum of the pin pair lengths is at the absolute minimum. For example, for a three-pin network, if the distance between pins A and B is 3, between B and C is 7, and between A and C is 5, the min-span pin pairs would be A to B and A to C. Incidentally, for a network of n pins, the number of pin pairs required to completely connect the network is always $n - 1$.

Sometimes electrical considerations result in additional constraints on network treeing. Two common treeing algorithms in this category are "daisy chain," in which each pin can only appear in at most two pin pairs, and "star," in which one pin (usually the network source) is paired with all other pins.

Most modern routers are not restricted to pin-to-pin connections. An extension to min-span, known as "Steiner treeing," is often used. As each pin pair is generated, it is treated as a "super" pin that is eligible to be paired with other pins. This allows the router to "tee" into existing connections rather than just going pin to pin. Using the previous three-pin network example, the Steiner pairing would be A to B and C to "super" pin A to B.

9.2.2   Ordering[3]

Most routers work on a single connection (pin pair) at a time. Once a connection is routed, it becomes a permanent part of the layout topol-

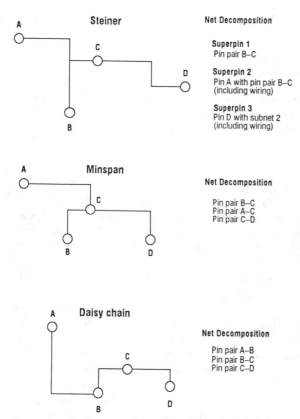

Figure 9.1 Interconnection alternatives. *(Courtesy of Personal CAD Systems.)*

ogy and is a barrier to further routing. Therefore, the order in which connections are routed can have a significant impact on layout results. As mentioned previously, a simple yet effective ordering is to sort connections from shortest to longest. Intuitively, this makes sense. As more connections are put down, the routing becomes more restricted. Since the possible paths of the shorter connections are more constrained than the paths of the longer connections, the shorter connections should be done first.

Some modifications are sometimes applied to this basic ordering scheme. One variation is to route all straight connections first. Another is to route congested areas first, such as around printed circuit board connectors. Often, higher priority is given to special signal classes, such as clock networks, in order to ensure that their routes tend to be as short as possible, even at the cost of decreasing overall router effectiveness.

9.3 Surface Organization[2,4]

In order to understand the way in which routers use wiring space, it is useful to define the types of space organization. These are gridded, gridless, plastic grid, channel, and combinations of these.

9.3.1 Gridded surface

The "gridded surface" method was first used for digitizing. The wiring surface is divided into a uniform grid of 0.010, 0.020, 0.025 in or some other convenient pitch in both X and Y directions. Wires can be placed on any grid line that does not cause an interference with a land, component lead, via, or keep-out area. Most routers are the gridded type; i.e., they center conductors and lands on an imaginary grid. These routers also rely on a strict set of design rules that determines the dimensions of the rectangular routing cell.

Gridded routers work best with designs that consist mainly of small-scale (SSI) and medium-scale (MSI) integrated circuits. The higher the component density becomes, the slower the router runs because it has to manipulate a larger number of coordinate vectors. Thus, the search and scan time for assessing each individual net increases as the component density increases. Another major weakness of gridded routers is that they cannot use all the available space during the routing process. However, a great advantage of gridded routers is that they are relatively easy to set up and to understand.

9.3.2 Gridless surface

Gridless routers also rely on design rules, but they are not constrained by fixed grid or cell dimensions. Instead, they assess trace width, conductor spacing, and via size for the net currently being routed, so they can vary the grid size on the fly. The resolution can be as fine as one-millionth of an inch. This technique allows stacking of conductors and spaces of different width across a span surface while dodging around obstacles such as pins and vias. There is actually a grid of very fine spacing. It is often used on two-sided surface-mount designs.

The gridless approach to board routing has been attempted by several companies, not always with a high degree of success. Design violations are often created when the board is laid out, and the placement rarely offers the degree of quality needed by an automatic router. Furthermore, the performance of a gridless router is inherently slow. The algorithm is computationally intensive, making extensive calculations each time a connection is made.

To understand this, contrast the gridless approach to a gridded costed-maze algorithm. With the gridded maze approach, the router does a quick expansion using simple integer mathematics. The gridless algorith, however, incorporates many calculations relating to encountered obstacles, optimum positioning, and design-rule considerations at each stage.

The major weakness of the gridless router is that after routing is complete, the board may need extensive manual editing. These routers may produce poor conductor alignment and nonstandard connections that may be difficult or impossible to manufacture. The greatest advantage of gridless routers is that they adapt easily to changes in topological and packaging technology, for example, the requirements of surface-mount devices.

When a printed circuit board has any significant size, the amount of memory and data-processing cycles needed to complete a route make it a poor choice. This is more often a choice when manual placement of wire is done. Manual editing is also difficult on a gridless system. Because the router does not include a grid, it is difficult for the user to complete the remaining traces which were not routed automatically without further violating design rules. Running the manual edits through a layout checker again impedes the design process, making the gridless approach the slowest methodology of all. Most importantly, gridless routing capability does not guarantee that a router will route to 100 percent completion with manufacturable results. If a gridless router can achieve 95 percent completion but takes 95 percent more time than a gridded router, nothing has been accomplished.

9.3.3 Plastic-grid surface[5]

Plastic-grid or flexible-field surface organization permits each routing grid to be of a different size. It is a recent innovation intended to deal with components that do not fit the same grid as the majority of the components. It may help eliminate hand routing of connections to off-grid components.

Run before the actual routing, the plastic-grid algorithm establishes a grid pattern of varying-sized rectangles which are customized to the components and other restrictions on the printed circuit board. As such, dense component placements and extraordinary land spacings, such as off-grid, are easily accommodated, with connections to the center of all lands.

The minutes invested in setting up the variable grid pattern are easily justified. The rest of the layout program is unhindered by the computational overhead that is incurred by other gridding approaches. In fact, for the 20 percent of the boards which cannot be routed with a fixed grid, the overall routing time is usually improved. In addition, a plastic grid improves the utilization of routing channels, resulting in boards with fewer vias or even fewer layers, which improves manufacturability. And should the user choose to route some traces manually, interactive editing is facilitated by the existence of the customized grids.

Both gridless and plastic-grid routing techniques are inherently slower than a gridded solution. Therefore, they are only useful on designs for which a gridded router runs into trouble, such as with fine-line designs and off-grid components. Because the plastic-grid solution builds upon an existing costed-maze algorithm, it may be implemented in such a way as to be user-selectable. It thus overcomes the inherent disadvantages of gridless algorithms.

Table 9.1 shows comparative performance of the same routing algorithms with a regular grid and with a plastic grid on three typical high-density boards. The statistics given are for the first comparative study made, and so do not reflect the highest possible performance differentiation. Rather, the standard performance improvement is illustrated. As is always true when selecting a CAD router, the best way to determine which solution fits particular requirements is to do a benchmark using real design problems.

It can be seen that the plastic grid is most effective with more demanding tasks. For example, in test C, a six-layer board was routed in about the same time, with about the same number of vias, by the two grid methodologies. However, the task was relatively simple and was completed in only 2 hours. When the same design was routed on four signal layers instead of six, a satisfactory completion level could not be attained by a fixed grid. However, the plastic grid achieved a 100 percent route in 27 hours.

9.3.4 Channel surface[6,7]

Channel-surface organization creates routing surfaces that are made up of a series of channels that pass between pins of the devices placed on the routing surface. Vias are placed in between the channels using the same pattern as the component leads. The effect is to prereserve space for both wire and vias, so that the placing of one does not diminish the space for the other. This is especially important when many routing layers are needed to hold the wire, as via assignment to resolve a wire in one wiring pair must not disrupt the space in another layer. This is the structure of choice for very dense printed circuit

TABLE 9.1 Fixed-Grid and Plastic-Grid Component Routing Performance

Board characteristics and routing results	Test A		Test B		Test C			
Components	287, surface mount 1 300-pin connector 1 165-pin connector		497, ECL		144, digital 4 pin-grid arrays 2 100-pin connectors			
Size, in	9–6		16–14		9–6			
Signal connections	1960		4712		995			
Density, in²/IC	0.18		0.45		0.39			
Conductor width, mils	5		8, 12[a]		8			
Clearance, mils	5		8, 12[a]		8			
Via size, mils	25		25		50			
Land size, mils	Varied		25		55			
Signal layers	4		6		6		4	
Power/ground	2		2		2		2	
Grid type	Plastic	5 mil[b]	8, 12 mils[c]	Plastic	20 mils[d]	Plastic[d]	2 mil[d]	Plastic[d]
Time to route, hours	16.7	22	23	15.2	2.75	2.2	35[f]	27
Completion, %	100	99.9	99	100	100	100	96[f]	100
Number of vias[e]	2112	2584	2071	1721	112	97	—[f]	389

[a]12 mils on cap layers, 8 mils on inner layers.
[b]Not all connections made directly to center of land.
[c]Cap layers and inner layers routed separately.
[d]Clocks prerouted manually.
[e]After auto cleanup.
[f]Router plateaus: no further nets are completed.
SOURCE: Cadnetix.

263

boards as it copes best with the high density of pins and the resulting lack of space for wire.

Channel routers, unlike the maze-running and line-probe routers, require a more rigid and disciplined component layout. They do an excellent job under ideal conditions. As the name "channel-router" implies, circuit components must be placed in such a way as to leave vertical and horizontal channels between the components through which connection paths may be run. The components must be placed in an ordered matrix providing for spaces between the rows and columns of pins that serve as the routing channels.

The basic channel-router algorithm operates on two-sided boards, making vertical connections in vertical columns on one side and horizontal connections in horizontal rows on the other. Any two points to be connected must lie in the same channel column or row for the router to make the connection. The path may meander within the channel to avoid barriers, previously routed paths, vias, and other obstructions. The meander is controlled to allow fixed amounts of horizontal movement in vertical channels and vice versa.

The router proceeds by connecting the shortest paths first and progresses until it has completed all the connections possible within a channel. It will then move on to the next channel and eventually to the other side. At this point, the basic channel-router algorithm is complete.

Most implementations go one step further by making use of a fixed- or floating-position vias. With this modification, the router will attempt remaining connections by changing sides during the construction of the path. The router continues to follow the basic channel rules during this process, i.e., vertical channels on one side, horizontal on the other.

The main drawback to channel routers is that they work best on a regular board geometry filled with open channels. These routers tend to be faster than Lee routers, however. And because the routes are not assigned until the end of the process, previously assigned traces can be moved aside easily to make room for new routes, as in rip-up routing.

9.4 Basic Routing Techniques[6]

The various routing techniques may be applied individually or sequentially as multistage routers. A good routing program attempts to make all the required circuit connections using the shortest total connection paths and simultaneously reduce the number of vias or optimum points at which the connection paths change from one layer of the board to the next to avoid obstructions. The process must be capable of constructing an intelligent routing pattern while constantly checking clearances and connectivity. All this must be one while keeping both routing and circuit costs to a minimum without sacrificing circuit quality.

Many automatic implementations are restricted to routing between pairs of layers using vias for feed-throughs, and can only be extended to multilayer configurations by concatenating layer pairs. This approach does not result in efficient multilayer routes. All layers must be taken into consideration simultaneously to accomplish this end. Most of the routers are grid-based, and as such can only make horizontal or vertical interconnection routes.

In limited cases, some routers allow diagonal interconnection runs. Some routers do not allow connection paths to run under components. Automatic routers are really the most effective for digital circuits, not analog circuits with many discrete components.

9.5 Routing Algorithms[1,2]

When discussing routing technology, it is important to understand the difference between routing algorithms and the implementation of routing algorithms. An algorithm is a foundation upon which an automatic router is built, and it determines basic router capabilities and performance potential. Implementation of an algorithm is equally important, however, and different implementations of the same algorithm can produce vastly different results. The parameters of automatic router operation that should be considered include route quality and speed, level of completion, and memory and processing requirements.

Most autorouters are based on Lee's algorithm, with the exception of a few vendors who use a line-routing algorithm. Line routers are generally fast, but are very inflexible and incapable of performing typical printed circuit board designs. Many variations to Lee's algorithm are used, with performance differences ranging from subtle to overwhelming. The differences in implementation of Lee's algorithm make it imperative that autoroutes be compared carefully to determine the performance and quality of routing for the type of design work being done.

9.5.1 Maze-running (Lee's) algorithm[2,6]

The maze-running algorithms are variously known as "exhaustive search," "wavefront," or "Lee's" routers. It is the characteristic of an unrestricted maze-running router to never fail to resolve a connection if a valid path exists. The router makes interconnections by expanding out from the grid primary connection point until the secondary point is found and connected.

The unrestricted maze runner will complete every possible path from the primary to the secondary component, compare all the successful connections, and pick the best (shortest path and/or least vias) available. It will repeat this exercise for all connections to be made. In essence, the maze runner consists of finding a conductor route to connect them whenever one exists. In general, this method requires considerably more processing time than the other methods.

The primary-user parameter employed is the size of the wiring rectangle defined. The smallest wiring rectangle is the area encompassed by the two points to be connected. The area can then be expanded, usually by grid increments, until it reaches the full routing area. Early connections will impact the quality of later connections with the maze router. Therefore, when using a multirouter system, it is best to use the maze router as the last step in interconnect, thereby minimizing the probability that an early route will block further routing capability.

However, sometimes the shortest path is not the best path. Some routers use a more sophisticated "costed"-maze scheme to better handle multiple layers. To best utilize routing resource, the router should prefer going horizontally on some layers and vertically on others. There is a "cost" associated with moving to an adjacent grid based on the direction of travel. A cost is also assigned for using a via. During fanout, each grid point is marked with the cost to reach it from the source. The maze algorithm will find the least-cost path (which may or may not be the shortest path).

A costed-maze router performs three basic operations. The first operation prepares the board for routing by partitioning the board into hundreds, or thousands, of cells. The size of each cell is determined by the routing grid (i.e., a 25-mil grid yields 25 × 25 mil cells). All the structures on the board (lands, copper area, traces, tooling holes, etc.) are marked in the cells in which they belong.

After creating the grid, the router selects the first land to be routed and begins the second process, called "expansion." The router examines all grid points in larger and larger concentric rings around the selected land until the destination is reached. To visualize this, think of the circular ripples that expand from the point where a rock is dropped into still water. As the router expands across the board, it assigns "cost values" to each cell in accordance with a cost table, Table 9.2.

A "cost table" is simply a matrix with cost values (or "weighting factors") assigned to a list of routing variables. A few examples of variables which may have weights assigned to them. For example, when

TABLE 9.2 Conductor Routing "Cost" Table

Description	Cost
Heading toward target using proper direction	1
Heading away from target using proper direction	3
Heading toward target using wrong direction	5
Heading away from target using wrong direction	8
Making a corner	+ 5
Adding a via	+ 9
Using cells next to any pads except target and origin	+ 7

SOURCE: Cadnetix.

expanding from one cell into the next, if the router were to move in the correct direction through the cell towards the target land, the cell may be assigned a value of + 1. Some weights may be assigned by the designer, while others may be fixed in the algorithm. In fact, it is not unusual for a cost table to contain 30 or more variables. Further weights can be added to the cost table to handle restraints imposed by the board design.

The router does not actually create a trace during the expansion operation. It only assigns costs. When the target land is reached, the third operation begins. A trace is laid from the target land to the origin land along the path with the lowest cost. Because a costed-maze router assigns all these costs before putting in a trace, it is able to use much more information than a line-probe router, resulting in higher-quality routes.

9.5.2 Line-probe (line search) algorithm[1,3,6]

Line-probe algorithms have significant performance advantages over the maze-running algorithms. Since they only keep track of line-end points, they do not require the large amounts of storage used by the maze runners. Line-probe routers also use a much simpler and more deterministic approach which requires significantly less execution time and therefore less cost.

The line-probe router connects a pair of points by constructing simultaneously a sequence of line segments (probes) out from each point to be connected. When the two-line-segment sequences intersect, the connection path is complete. The drawback to this method is that the line-probe router, unlike maze runners, will be unable to route some connections that are routable. However, there is a relatively high completion rate for line-probe routers, usually over 95 percent.

A line-probe router is capable of a number of decisions, including changing direction, addition of a via, and backtracking. Theses routing decisions are prioritzed in the algorithm to enable it to find the most efficient path between lands. By continued repetition of these steps, the trace reaches the target, using priorities to make correct routing decisions.

One way to achieve greater speed with the router is to start a trace from two lands simultaneously, extending the two traces toward each other until they meet. Features also can be added to make an algorithm function more efficiently. Instead of making a decision when it runs into an obstacle, a line-probe router can be programmed to stop periodically to evaluate its route, then to make a decision based upon the information it has gathered. There are many variations to this approach. Most routers use more sophisticated methods to generate escape points. Another common technique is to shoot out lines from both the source and the target and attempt to meet somewhere in the middle.

The major advantages to the line-probe technique are its speed and the straighter paths that it produces. Its major disadvantage is that it sometimes misses valid routes, particularly in the latter stages of routing when the surface is crowded. There are some routers which combine both the line-probe and maze approaches.

9.5.3 Channel-routing algorithms[6]

The channel-routing algorithm and its many enhanced versions is second in implementation popularity only to the maze runners. Channel routers, unlike the maze-running and line-probe routers, require a more rigid and disciplined component layout. They do an excellent job under ideal conditions. As the name "channel-router" implies, circuit components must be placed in such a way as to leave vertical and horizontal channels between the components through which connection paths may be run. The components must be placed in an ordered matrix providing for spaces between the rows and columns of pins that serve as the routing channels.

The basic channel-router algorithm operates on two-sided boards, making vertical connections in vertical columns on one size and horizontal connections in horizontal rows on the other. Any two points to be connected must lie in the same channel, column, or row for the router to make the connection. The path may meander within the channel to avoid barriers, previously routed paths, vias, and other obstructions. The meander is controlled to allow fixed amounts of horizontal movement in vertical channels and vice versa.

The router proceeds by connecting the shortest paths first and progresses until it has completed all the connections possible within a channel. It will then move on to the next channel and eventually to the other side. At this point, the basic channel-router algorithm is complete. Most implementations go one step further by making use of fixed- or floating-position vias. With this modification, the router will attempt remaining connections by changing sides during the construction of the path. The router continues to follow the basic channel rules during this process, i.e., vertical channels on one side, horizontal on the other.

9.5.4   Heuristic (pattern-recognition) algorithms[1,2,4]

Heuristic routers are the simplest type of router, and were the first to be used in CAD systems. A heuristic, or "pattern-recognition," router is one which is written to solve a single, specific design problem. The algorithm typically completes a connection in a fixed manner once a particular set of predefined requirements is met. Heuristic routers are thus often referred to as "dumb" routers, due to their limited capabilities.

However, heuristic algorithms still have their uses. For example, they are commonly employed to route memory arrays, where each pin is connected to the equivalent pin on the next component in line (i.e., pin 2 on U1 will be connected to pin 2 on U2, U3, U4, etc.). Because the familiar 45° pattern is consistently used to route memory arrays, it is easy to develop a heuristic router that recognizes memory arrays and generates the regular 45° trace pattern.

The fact that heuristic routers only can route when confronted with a specific pattern of pin placement and connectivity is not necessarily bad. When these criteria are fulfilled, the trace pattern used usually is the most efficient way of making a connection—given the limited information that the router can manage. In addition, heuristic routers usually are very fast because the router needs only to search the board for predetermined configurations, then drop in the appropriate trace pattern, without having to execute a complicated mathematical algorithm for each conductor.

When the pattern router is used in conjunction with other routers and restricted to routing the repeated patterns, it can save much time. However, if used alone, it can create such a confusion of traces and vias that nonrepeating nets cannot be completed, and it may then be necessary to reroute the whole board. Also, if just one component is moved up or down by even 100 mils, the pattern does not meet the criteria that have been set for the router, and consequently, the pattern is not "recognized" by the heuristic algorithm.

9.6   Routing Implementation[2,7]

A sophisticated router is implemented in a way that enables it to store and use much more information, resulting in a router that is better

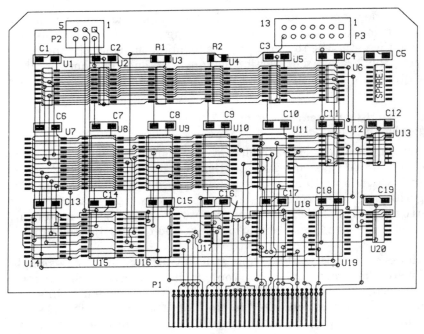

Figure 9.2 Routed layout. *(Courtesy of Personal CAD Systems.)*

equipped to handle unique technologies and more difficult routing problems (Fig. 9.2). In short, a sophisticated router better approaches the skill level of experienced designers and their ability to work with large amounts of information.

With early automatic routers, the information in the database was purely graphical and contained only limited connectivity information. Consequently, routing was purely a graphic problem. If the router encountered a via, land, or trace while trying to connect two lands, all of these objects were treated the same—as obstacles in the way of a connection. It had no information with which to discern the differences between objects on the board. To make a router more "intelligent," more information must be contained in its database. Using the electrical information provided by a schematic, the router can be programmed to make more informed routing decisions. Copper sharing is an example of a quality improvement that can be gained by using more information from the database.

This concept can be extended. The information necessary for copper sharing already existing in the schematic only needs to be included in the routing database. But there are other types of routing information that are specified at the beginning of the routing process which could be combined with schematic information to give additional capabilities to the router. Examples of information created for routing include conductor width, clearance, via size and shape, selected layers to be routed, maximum allowable connection length, and routing priority. When combined with signal information, emitter-coupled logic (ECL), transistor-transistor logic (TTL), power/ground, etc., from the schematic, the automatic router can operate on each type of signal using appropriate design rules. Routing can be done simultaneously, eliminating the multiple passes needed to change design rules. This not only saves time but also produces better routing results.

9.6.1   *X-Y* routers

The X-Y router routes wires in only one direction on each routing layer. This means that wires that are not purely horizontal or vertical always need at least one via to complete them. In simpler printed circuit board designs this adds cost. In more complex designs, this method results in better utilization of each wiring layer because early wires are all parallel in a given layer and do not interfere with later ones.

9.6.2   Orthogonal (cut-based) router[7]

Orthogonal routing gets its name from 90° bends in routed conductors. The procedure plans a conductor by proceeding in a straight line until an obstacle is encountered. Then, the router creates a via (plated through-hole) to get to another layer of the circuit board. Routing continues on the new layer in a direction 90° away from the original trace. The software continues to turn corners and make vias in this manner until it either finds the destination or gives up trying.

Routing software on personal computers makes most general signal connections on printed circuit boards with orthogonal routing. But these algorithms have a few fundamental limitations, particularly when used on small machines. One limitation pertains to board layers. Because orthogonal routers create vias when they change direction, the software cannot route in a single layer; the board must contain at least two layers of traces. (Memory and bus routers generally work on a single layer.) This sort of routing is sometimes called "cut-based," because each printed circuit board layer is sectioned into cuts or slices in one direction.

Other restrictions concern multiple-layer boards. Orthogonal routers found on small computers generally route on only two layers at a time. Thus, software would treat a six-layer board, for example, as a sequence of two-layer routing problems. Tables relate current layers with connections and vias already made on others. Dividing the routing in this manner may reduce the chances of successfully completing complex multilayer boards. Additional qualifications concern connections made between routing layers. Some packages cannot create so-called buried vias (junctions between buried layers). In particular, lack of such facilities can make it difficult to design boards containing surface-mounted devices (SMDs).

9.6.3 Rip-up and reroute router[4,5]

Several routers have been developed to cope with densely populated circuit boards. These have been known to complete 99 percent or better of the connections on complicated boards. Perhaps the most well known of these routers is called "rip-up and reroute." Rip-up routers get their name from their treatment of routed connections. Basically, the algorithm modifies paths discovered early in the routing process to make room for later traces.

Rip-up routers make several routing iterations. After each pass the program examines traces that block paths to unconnected pins. On succeeding passes, the program reroutes the blocking traces and attempts to complete the rest of the connections. In contrast, ordinary orthogonal routers generally attempt to route boards in one pass, and never change routes placed on the board.

The rip-up (multipass) router attempts to complete 100 percent routing of a board by automatically performing many iterations of the routing process. At some critical point in the process, the rip-up router may defy some of the design rules in order to complete the routing; in that case, when routing is complete, the router performs another iteration in an attempt to clean up the areas where it broke the rules.

Because this type of autorouter requires a large database, it should be used on a computer with a large memory (as mainframe or a workstation) in order to reduce the number of disk accesses. Further, because the rip-up router must perform a very great number of computations, it may be slow unless you run it on a large host or a workstation that is equipped with a hardware accelerator. The rip-up router also tends to create an excessive number of vias during the later iterations. Therefore, it expends most of the available running time on reviewing and removing vias.

The main difference between maze routers and rip-up routers is that a rip-up router will continue to "rip-up" or move existing routes in order to place new routes until a 100 percent completion rate is achieved.[8] This may take hours, days, or in some cases weeks. In contrast to this, a maze router will not disturb an existing route and will only route a new trace if it can locate a path that has not been previously blocked. This means that maze routers may not achieve a 100 percent completion rate, but the run time tends to be only minutes or hours rather than days.

Maze routers allow for more interactive usage and for greater control over the routing parameters than rip-up routers. They also provide for reentrant capability, meaning that they can be stopped and restarted after making edits to the database. Rip-up routers are typically used for small to medium-sized boards where trace length and aesthetics are not the biggest concern.

9.6.4 Squeeze-through and shove-aside router[9-11]

The squeeze-through and shove-aside router uses a variation on gridded routing that permits two widely spaced traces to pass between two pins by drawing them together as they pass through the land area. They are used to stretch the usefulness of gridded routers in denser designs. Although the iterative routing techniques are of great value in a discretionary via site environment, they tend to falter on fixed via site designs. A considerably more challenging routing problem arises when vias can no longer be freely introduced to unravel wiring crossovers; iterative perturbations of previously established wiring must be performed with greater cunning if acceptable routing yields are to be attained.

The squeeze-through and shove-aside router is really two separate algorithms that work well together in many autorouting situations. The purpose of squeeze-through is to create routes in a topologically correct position with respect to existing paths and vias. This is accomplished by routing with free grids when possible, and using grids too close to other wires and vias when needed. These clearance violations are penalized by a small cost. However, crossing centerlines of other wires is disallowed so that the topological correctness is guaranteed.

The squeeze-through router is capable of shoving aside entire bundles of previously routed leads to open up feasible wiring paths for additional interconnections. If the shove procedures are unable to make room for an interconnection, it reverts to the iterative rip-up and reroute strategy. The power of the squeeze-through method derives from the fact that one shove maneuver can accomplish what would otherwise require a lengthy sequence of rip-up and reroute cycles (see Fig. 9.3).

Using an autorouter based on interactive squeeze-through and shove-aside techniques can save time and money. All too often people evaluate autorouters solely on the basis of completion and time. While these are important, they do not tell the whole story. When an

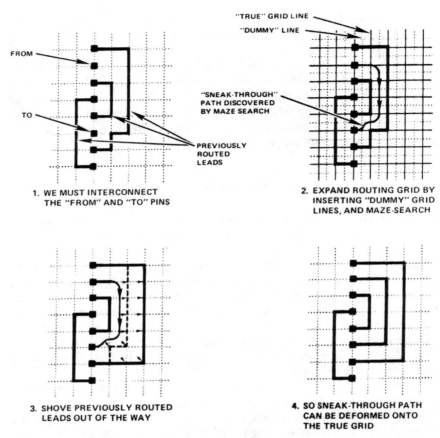

Figure 9.3   Squeeze-through conductor routing.   *(Courtesy of Automated Systems.)*

autorouter does not reach 100 percent completion, how long does it take to finish routing the board manually? That clean-up time ties up a designer and workstation. The designer is wasting time shoving leads and cleaning up routes to make room for the last routes. These are generally bad routing that should never have been put in in the first place.

If a board cannot be autorouted, for whatever reason, how long does it take to design it manually? This is when the interactive capabilities made possible by a shove router can save the most effort. The tedious work of moving and rearranging routes is best left to a shove router.

9.7   Postroute Processing[3,12]

Once the routing is completed, there are a number of postroute processes that can be applied to enhance the manufacturability of the design. One example is via minimization. A simple approach is to look at a path between two vias, or between a pin and a via, and see if it can be moved to a different layer. A somewhat more sophisticated approach is to look at pairs of paths and see if it is possible to simultaneously move them to different layers to eliminate vias. Another possibility, once the board is finished, is to attempt to reroute each connection to see if it can be done with fewer vias.

Another postroute process is a beveler which changes 90° corners into 45° angles. (Note that only a few of the more sophisticated routers actually allow the maze algorithm to expand in a diagonal direction.) Still another example is a process which spreads features apart to increase clearances between items.

9.8   Router Performance[9]

Router performance is measured from two perspectives: economic and technical. The economic factors are simply based upon comparing the cost to produce the circuit with a particular router method to manual or other routing method costs. The cost includes the resources necessary to complete the layout as well as the production cost for the circuit. Technical performance is somewhat more difficult to measure because it compares the quality of the finished circuit in terms of signal degradation, noise, crosstalk, and electronic reliability.

The performance of a routing algorithm is measured against three criteria: percentage of routing completed (or conversely, the amount of manual effort to complete routing of the layout), execution time, and system storage resources required. The best algorithms balance all three criteria. Autointeractive graphic systems that include a routing subset must allow the designer to interact with the system in order to guide and complete the routing process. If the result of the router is less than 100 percent completion, facilities must be provided by the system that allow the user to attempt to finish all incomplete connections.

The system must balance the capabilities of the automated routing subsystem with effective graphic editing functions. The system must provide an adequate human-machine interface that allows the user to complete the connections while the system monitors the user's actions against the original ground rules for interconnection, and signals the user when any of the rules are violated. With these facilities, the user can develop connections based on experience and creativity that are beyond the capabilities of the system alone.

A reentry capability in routers is desirable. Reentrance means that the user can stop the routing procedure at some intermediate point, inspect the results in a graphic form (usually presented on the CRT), manually edit changes in physical positions, and restart the routing procedure. The manual changes can involve movements or deletions of conductor lines or components. Re-entrant capability is often used in making changes on previously laid-out printed circuit boards.

Some routers have a preroute step that displays a "rat's nest." This helps the user identify potential routing problems. The user can then change the placement, choose one of many different routers for the job, or set parameters appropriately. Designers will require significant training and practice to become proficient at systems with both automated routing and user involvement capability. It may be months before predicted time and cost reductions are obtained.

There currently is no router that can route every board 100 percent every time. Most designs beyond minimum complexity will require manual routing when the automatic router is finished. These last routes of the design are the most difficult to connect, and may necessitate changes in placement, routes, or layup. This is often the most time-consuming part of the design after placement.

References

1. Robert Anastasi, Cadnetex Corp., "Survey of PWB Trace Routing Technologies—Part 1," *Electri-Onics*, August 1987, pp. 21–23.
2. Robert Anastasi, Cadnetex Corp., "Survey of PWB Trace Routing Technologies—Part 2," *Electri-Onics*, September 1987, pp. 42–44.
3. Robert Dean and Howard Schutzman, "A Layman's View of Placement and Routing Algorithms," Personal CAD Systems, Inc., San Jose, Calif.
4. Aptos Systems Corp., "Checklist for Comparing CAD Systems and Software," *Printed Circuit Design*, January 1987.
5. Robert Anastasi, Cadnetex Corp., "Survey of PWB Trace Routing Technologies—Part 3," *Electri-Onics*, October 1987, pp. 26–28.
6. "Automated Design Guidelines," IPC-D-390, Revision A, February 1988, Institute for Interconnecting and Packaging Electronic Circuits, Lincolnwood, Ill.
7. Leland Teschler, "Fresh Look at PCB Routing," *Machine Design*, May 22, 1986, pp. 77–82.
8. "Printed Circuit Board Design—Illustrated User's Guide," 1989, Personal CAD Systems, Inc., San Jose, Calif.
9. Henry C. Bollinger, Automated Systems Inc., "A Mature DA System for PC Layout," *IPC Technical Review*, November 1977, pp. 10–17.
10. Kenneth R. Wadland, Engineering Software, Inc., "The Importance of Shove Routing," *Printed Circuit Design*, October 1988, pp. 20–25.
11. John Cooper and David Chyan, "Autorouting Today's High Density PCBs," *Printed Circuit Design*, October 1988, pp. 36–46.
12. Howard Schutzman, Personal CAD Systems, "A Behind-the-Scenes Look at Autorouting," *Printed Circuit Design*, December 1988, pp. 34–40.

Design Checking and Postprocessing

10.1 Introduction[1]

The advent of automated systems used in the design process has both simplified and complicated the checking process. In manual designs everything was checked, because everything could be a possible source of error. Component sizes, hole sizes, conductor clearance, conductor widths, land-to-hole ratios, board areas, electrical interconnections, positional accuracy, etc., were all reviewed with a great deal of care. Automated design requires the same care. Someone once said that "Automated is the fastest way we know to generate scrap." And so the task of checking becomes doubly important. Some of the reasons designs are checked are

- CAD systems are not infallible.
- Mistakes can be made.
- Safeguards can be circumvented.
- Engineering changes may not be updated throughout the data.
- Tooling must be able to provide the required part.
- Designs may not represent an optimum solution for a producible part.

Many types of checking are performed during the design cycle. Some are done by the electrical or mechanical engineer before, during, or after a design is complete. Some checking is done by the board designer, mostly during or at completion of the design cycle. Another series of checks is made by the process engineer, whose responsibility it is to see that the product being manufactured matches the documented requirements. Finally, the systems engineer checks to see that the part works in conjunction with other parts.

10.2   In-Process Layout Checking

Most design layouts start with the receipt of a schematic or logic diagram in rough-draft form and a series of notes that represent the board to be used and any design restrictions. This information is entered into the computer via punched cards, teletype, keyboard, or menu commands. In addition, references to a computer library, or input from another program, result in the creation of the design layout database. If this information is not correct, it it does not represent the design information in all aspects, or if the output parameters produce data that violate manufacturing standards, then one should not proceed with the design layout.

Checking input data is most important to the design layout process. For instance, regarding net lists, some companies have two individuals separately encode the electronic diagram and enter the data into the computer. Although different code numbers and reference designators may be assigned, the computer can still compare the interconnectivity of the two inputs and usage of library parts. Differences between the two inputs are printed out and resolved by the designer.

Another checking technique is one that compares the interconnectivity of a schematic and logic drafting automated system with the encoded input of the electronic diagram. The differences are graphic symbols versus actual parts; however, the checking program understands the association between the various elements and also can compare the interconnection between the graphic and encoded input. Possible errors are printed out and resolved.

10.3   Cleanup and General Checking

When printed circuit boards are generated with component features such as lands and/or conductors that do not fall on a grid center, a cleanup program can physically move conductors to "center" them between lands and thereby reduce the possibility of spacing violations. In general after completing the cleanup function, one should check the following:

- Input checks
    a. Board outline dimensions
    b. Net list data
    c. Part types, shapes, sizes
- Process checks
- Design rule violation checks

    a. Conductor-to-conductor spacing
    b. Conductor-to-land spacing
    c. Land-to-land spacing
    d. Land-to-via relationships
    e. Via-to-via relationships
    f. Isolated copper
    g. Mechanical clearance
    h. To edge of board
    i. Around on-board connectors
    j. Around mounting holes
- Project standard checks
    a. Manufacturability
    b. Identification
      (1) The printed circuit board number
      (2) Revision block
      (3) Master pattern number
      (4) Revision number
      (5) Assembly title and number
      (6) Serial number
- Physical checks
    a. Continuity Check

Many CAD systems have the capability to do many of the design checks. Connection and dimension checks are generally a part of the CAD software package. Net-list-driven CAD systems usually have the capability for circuit continuity versus layout check. A program to check the physical layout to the schematic net list is desirable. Figure 10.1 is a typical checklist that should be completed to verify the design.

Design rule violations, conductor widths, spacing, land and conductor spacing, etc., can be caught before board production. This is provided by a design rules checking program. Samples of errors caught are shown in Fig. 10.2.[2] Once the design rule errors have been fixed, the layout is ready for production.

Design rule criteria resident in an existing system may not be fully compatible with the introduction of a new technology such as surface-mount or hybrid designs. Design rule checking means two things: continuity verification of all signal and power pins, and clearance violation checking. Clearance checking is more vital than ever due to the tight tolerances used on technologically advanced designs. The system program must be able to check different clearances on different sections of the board. A feature which allows the designer to define any given area or layers and perform user-defined gap checking on just those areas is highly desirable.

As far as continuity checking goes, single-sided pad mounting pre-

ROUTINE CHECK LIST

DWG. NO._____   DATE_____   ENG._____

MODEL_____   EN.NO._____   STARTED_____
                                                    COMPLETED

1. Preliminary:
 ____ a) Log in and record job by En. No. and Charge No.
 ____ b) Check for the following support data, such as notes, layouts, sketches, reference prints,
      library shapes, etc.

2. Schematic
 ____ a) Check schematic symbols for correct nomenclature and type.
 ____ b) Check Schematic against engineers marked print, or net list.
 ____ c) Check schematic reference designations against assy, silkscreen and parts list.
 ____ d) List all spare gates and unused component designations.
 ____ e) Check for proper text and format (notes, etc.).

3. Artwork (Check Plot)
 ____ a) Check placement of components, proper size, center to center distance and pad size
      against parts list.
 ____ b) Check circuit continuity against schematic, including power and gnd (separate layers for gnd and
      power if multilayer).
 ____ c) Check for shorts or any other violations (mechanical and electrical).
 ____ d) Check for correct location and proper board identification and revision level.
 ____ e) Check for proper grid utilization.
 ____ f) Check registration and reference dimension.

4. Fabrication (Drilling Dwg.)
 ____ a) Does the physical size meet the mechanical requirements?
 ____ b) Are all the hole dia. the proper size (plated or non-plated)?
 ____ c) Off grid components are to be dimensioned.
 ____ d) Check for proper dimensions, tolerances datums and tooling holes.
 ____ e) Check that artwork features are properly located to datums.
 ____ f) Check for proper notes and format (notes, etc.)
 ____ g) Check for proper locations of UL logo and vendor date code.

5. Solder Mask
 ____ a) Check component and solder side for proper registration.
 ____ b) Check for proper clearance around pads. (0.015 to 0.020)

6. Silkscreen
 ____ a) Check for proper reference designations, locations and text size.
 ____ b) Check for correct location and proper board assy. identification.
 ____ c) Check polarities.

7. Assembly
 ____ a) Check for proper size and shape of components.
 ____ b) Check for component designations, polarities,or any other designations required.
 ____ c) Check for correct identification of board no.
 ____ d) Check for proper text and format (notes, etc.).
 ____ e) Parts lists
 ____ f) Check any miscellaneous mechanical parts and hardware (ejectors, heat sinks, etc.)

8. Master Artwork
 ____ a) Check in accordance with IPC-D-310.

Figure 10.1 Routine checklist. *(From IPC-D-390A.)*

sents a whole new problem for existing programs written to deal with continuity verification on through-hole designs with through-hole vias. Another problem inherent in CAD hybrid design is the occasional necessity to superimpose a part of the design in the post-processing/output stage. By definition, neither continuity checking nor clearance checking has actually occurred between the superimposed files or layers; only portions of the board will have been checked.

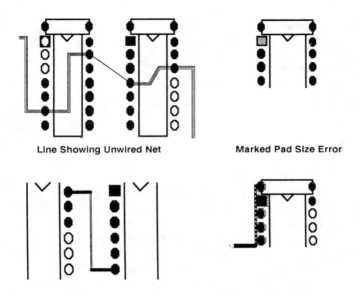

Line Showing Unwired Net   Marked Pad Size Error

Marked Trace Size Error   Marked Trace Spacing Error

Figure 10.2 Design rules violations. *(From IPC-D-390A.)*

Because work-arounds are often necessary to complete a high-technology design on a CAD system, automatic design rule checking often cannot be used. Manual checking is often necessary, along with detailed notes on all work-arounds.

10.4 Library Checks

Most automated systems use a library for the storage and retrieval of data that are in common usage during the printed circuit board design phase. The information contained in a component library has usually been checked and rechecked many times. Library integrity should be protected in such a manner that only the official librarian is permitted to change the information in the library. A problem can occur when a component is required that is unique to only one design and it has been decided that the information should not be a permanent part of the library. A typical example of this would be the need for a special transformer or coil or a particular board assembly.

The type of information required for the various elements listed in the library might be as follows:

- Overall component real estate required.
- Grid displacement of component

- Land-mounting requirements
- Hole-mounting requirements
- Blocked areas for no conductor routing
- Graphic electrical symbol (schematic)
- Graphic mechanical symbol (assembly drawing)
- Pin designation
- Board legend/nomenclature
- Electrical Characteristics

Some companies require that a single individual approve all new components prior to the start of a design. Thus all data entry about the component is performed and checked prior to submitting the package to design. Another technique is to allow the designer to enter all data including a new component at the time of design, and then have the computer "plot back" the information in a form that can be readily checked. (Lands compared with hole size, stating minimum annular ring left; lands compared with component outline; blocked areas compared with component outline; conductor routing paths compared with land position; blocked areas and component outlines; etc.).

If special hardware is required with the part, it must also be evaluated and checked. In fact all items that impact the relationship of the new part with existing standard parts must be evaluated.

10.5   Corrections

Engineer changes and design corrections must be approached with the utmost care when designing on a CAD system. Any changes made to the design must be implemented in all affected parts of the board database. For example, if the engineer changes a signal on the schematic, therefore resulting in a layout change, that change must be reflected in the net list and logical schematic database as well as the printed circuit board layout database. Corrections and changes at the end phase of the design are a breeding ground for error if not incorporated into the database and then checked thoroughly. Changes should be documented per revision level in accordance with validation phase requirements.

10.6   Back Annotation[1]

Once a design has been completed, all data must be consistent within the design description database. During the initial schematic capture, gates were assigned designators that were later changed in the design process. Gate swapping or combining into packages must not be incorporated into the released schematic or logic diagram. The designator for the gate or logic function on the diagram must be back-annotated using the part designator assigned during the part placement portion of the program. Thus, if a part is assigned "U12" as a designator, the first logic function would be U12-1 or U12A. Subsequent gates are assigned the next number or letter, whichever is conventional.

10.7   Postprocessing[1]

Postprocessing is the term used to describe the "after-design" steps that are required to produce artwork masters and other printed wiring tooling. Computer programs can be generated to enhance and convert data generated on various systems to a common interface language that can be used to accomplish the desired results. The postprocessor, to function effectively and meet the goals for which it is designed, must be directed to those areas where quality control tends to slip and/ or where computer standardization presents definite cost reduction advantages, such as

- Generation and control of engineering drawing and artwork
- Tooling
- Manufacturing and testing
- Quality

The postprocessor, when used with coordinating techniques, can take over and perform all the various tasks that are usually accomplished by humans when the artwork of a printed wiring image is contacted, tooled, balanced, and manipulated into the final artwork for a standard manufacturing panel. Tooling holes, test coupons, etc., are added to the board database in preparation for CAM postprocessing. CAD programs such as NC drill plot verification produce a drill map of all drilled holes at 1:1 scale. This plot can be used later by manufacturing quality control.

The printed circuit board postprocessor converts data generated by the CAD system into data that will set up and produce plot tapes for the engineering prototype, manufacturing artwork, and NC tapes for manufacturing equipment.

10.7.1   Terminal requirements

An interactive CRT terminal, with keyboard and cursor, should be installed in the computer graphics lab to input the computer graphics and manufacturing data. The installation should also include a CRT copying device that can be used to copy various machine setup data and other data required for record keeping. The terminal should allow the operator to answer all questions posed by the postprocessor by means of the CRT cursor or the CRT keyboard. This includes adding or deleting tooling and manufacturing features to and from the artwork.

The terminal should also be capable of displaying a single-line diagram of the manufacturing panel in simple form; that is, outlines of the panel, printed wiring images, and test data presented in proportional size to each other.

10.7.2   Program description

The following features are necessary for the postprocessor program.

- The postprocessor program should contain all steps necessary to separate photos and other printed board tooling. In its ultimate form, the postprocessor may be capable of determining the size of a manufacturing panel and positioning different layers of a printed circuit wiring board onto the panel, then adding test coupons, drill setup lands, plating bars, revision level, and data of the artwork. It may also produce an NC drill tape coordinated to match the panel.

- The processor should contain a library of tooling and manufacturing data. These data should consist of test coupons, indexing cross hairs, etc. Coding for new or special modules will be keyed into the library from the computer graphics terminal. The program should display these in true-width form to assure correct coding.

- The standard aperture setup should be provided. A capability of defining shaped lands must be added to satisfy the CAD input requirements. A means of changing the aperture setup should also be included.

- The standard packed character set should be provided.

- The present zero offset and photoplotter offset should be in the processor along with the capability of changing them.

- In the hard-tooling operation for a double-sided board, the processor should select the data in the last layer file of the data set.

- All automatic drill setup routines must be in the post processor.

- A means of expanding the manufacturing panel in order to provide space to add test and tooling features should be provided. This should be accomplished in the same manner as the board limits dimensions are varied. However, since the board limits data are furnished by the CAD system, the operator should be able to access and modify its size, either by subtracting an amount from the lower or left side of the board limits or adding an amount to the upper or right. These points are identified in the hard-tooling logic. This change will take place after the program has generated a normal dual panel and displayed it on the CRT.

In order to give the operator a feel for space, the dimension for both the panel and the board should be displayed, when requested, along with a border grid that surrounds the outline of the artwork panel. A feature which allows the operator to place the cursor cross hair on the edge of the board limits outline he or she wishes to expand and to type in a single command such as the following is extremely beneficial

ADD N.N INCHES

The operator may continue to change the other three edges and also make other additions, changes, or deletions before asking for a redraw. To obtain a redraw, the operator will key in the following:

REDRAW

The processor should respond by clearing the screen, generating the new panel, and then return by displaying it on the screen. The operator may repeat the operation if more changes are required.

- A means of calling, placing, or deleting the manufacturing data should be provided. As much of this input as practical should be done via CRT cursor and screen menus to reduce terminal time. Most fixtures would be located to the nearest 0.100-in accuracy (at 1:1 scale) with the same point being used by the processor to place related data on other layers of artwork. A CRT border grid that surrounds the artwork should also be included to aid in the operation. The processor prevents the operator from placing any feature within the CAD board limits perimeter.

- The processor should be able to determine if all apertures are available to plot the artwork. The routine must be capable of searching the data set files, finding the layers that contain copper planes, determining the land diameters and circuit widths used on those layers, adding twice the space allowed between the copper plane and circuit, and adding these to the width table. Then, the routine should compare the modified width table to the aperture setup, and return to the processor with a list of the aperture positions used and a list of apertures that are missing.

10.7.3 Input to processor

Input to the postprocessor will be from a data set created by the CAD system and should contain coordinate data to locate all circuit features (lands, conductors, etc.) that are required to make each printed wiring layer and all marking complete. The data will be stored on a data set and may consist of the following.

10.7.3.1 Parameters. The stored data should contain the following parameters to set up the postprocessor:

- Part number (detail)
- Board limits—two points (lower left/upper right, x, y coordinates) which define the smallest rectangle that would completely contain the board.
- Width table (land and conductor size)
- Hole table (finished hole sizes with plated holes flagged).
- Flags indicating each of the following:
  (a) Layer of the board
  (b) Flag indicating if the layer contains a copper plane.
  (c) Minimum space between copper plane and circuit.
  (d) Total square inches of copper and gold on outer layers.
  (e) Type of layup if multilayer.
  (f) Flag indicating a separate marking (silk screen)
  (g) File if provided.

Note. It would be advantageous to include a charge number and a flag to indicate whether prototype artwork or production artwork is to be generated.

10.7.3.2 Board database

There must be a file for each layer of the board starting from the top (component side) down. Each file must contain the following data, optimized to reduce plotter run time.

- Land data, giving X, Y coordinate location and size, including two dimensions if shaped.
- Line (circuit), giving width, number of points, start point, turning points, and end points.
- Rectangle data, giving pair of points defining a rectangle of flashable size by the window aperture (i.e., 2 in), and a flag indicating whether it is part of the copper plane or circuit.

If marking (silk screen) is required, the file containing that data (usually line data) would follow the last artwork layer file.

The last file of data will contain the information needed to drill the holes (presented from the component side). These data must be optimized to reduce drill movements. The following data are required:

- X-Y location of each hole.
- The finished size of the holes.
- Whether the hole is plated or not.

10.8   Line Drawing Procedure

The line drawing procedure generates pen plots of the printed circuit board. Plots are generated for the component side, solder side, and any unrouted connections. The primary purpose of the procedure is to produce drawings to aid in the manual completion of board wiring. In addition, plots can be used to aid in the verification of component descriptions. The properties of the CAD output for a line drawing may be summarized as the

- Board outline
- Board assembly number
- Revision code
- Blocked wiring areas
- Terminal areas (dummy, used and unused)
- Wires drawn on grid (in the same positions as they appear on the database)

10.9   Step and Repeat

In order to better utilize manufacturing-size panels, printed circuit board images and all associated information (hole sizes, hole locations, etc.) can be stepped and repeated to attain maximum utilization of the panel. Some CAD systems provide step and repeat programs. Step and repeat can also be accomplished with custom software or photographically. Drilling printed circuit board panels usually occurs in a format called a "cookie sheet." Figure 10.3 depicts a cookie sheet and shows the number and spacings of the various sizes of printed circuit boards that can be superimposed on it.

The pattern of holes for a cookie sheet are derived from the pattern of holes for a single printed circuit board designed by the system. Thus, if the system is used to design a double-sided printed circuit board, four identical hole patterns are needed for the cookie sheet, i.e., the hole pattern for the double-sided printed circuit board is stepped by a defined increment in X and Y and repeated four times to effect cookie sheet drilling. Since the spacing of the steps and the number of

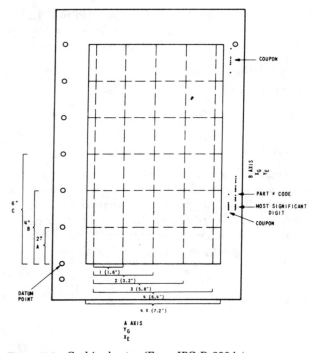

Figure 10.3 Cookie sheet. *(From IPC-D-390A.)*

repeats depends upon the printed circuit board size, the "drilling" program must be informed (by means of parameter data) of these values for a given design.

10.10   Rotation and Mirroring

Rotation and mirroring should also be controlled for all step-and-repeat operations. The following rules will provide a means of using one CNC/NC tape to step and repeat artwork for all layers of a printed circuit board, allowing the simplest operation in the photographic laboratory.

The data for each layer of artwork received by the tooling processor are looked at through the printed circuit board from the top (component) side. The text for each layer is already mirrored in place by the main processor (or CAD) system, each layer is handed to the tooling processor in a separate file starting with the top layer (component side) and continuing down through the board for each layer following the layup order of the printed circuit board.

10.11   Output Data

At the conclusion of the photo drawing procedure, the following drawings (expressed in terms of plotter commands) are normally recorded in the order given in the plot database.

- Photomaster for component side
- Photomaster for solder-side mirror image
- Negative for component-side ground plane
- Negative for circuit-side ground plane mirror image
- Board assembly drawing including board outline, component outlines, and reference designators
- Board assembly reference designator tables (optional—none or one drawing per table)
- Dot pattern
- The drill edits
- Paper tape listings

Standardization of output data is desired if designers intend to share the information with other systems. CAD is integrating with CAM/CIM throughout industry, i.e., printed circuit board manufacturers can now use CAD data to assist in generating production-ready phototooling. Designers should limit output formulas to industry-accepted formats such as IPC-D-350.[3]

10.12   Drill Data

Postprocessing programs are used to run media-generating equipment for processing of holes (drilled/punched) in the printed circuit board. This is most commonly in the form of drill tapes that are used to select and position NC drills. However, hard copy of X-Y locations of holes can be used to fabricate drill templates by manual or automatic methods. NC drill output by the CAD system's postprocessor should optimize the tool selects and X-Y locations of holes to minimize tool travel.

These same coordinate location listings can be used on various types of inspection equipment. Database information can generate hole size and location drawings for use in fabrication and inspection of the finished printed circuit board. Hole sizes are coded by differently shaped symbols, and are defined in a hole chart in the field of the drawing. It is advantageous for the postprocessor to produce a setup report along with the part program for ease of loading tools and special operations required to drill the panel.

Systems available today can also be adapted to a "paperless" format where computer data can be directly accessed by the printed circuit board fabrication operation and the information transferred by a direct line or magnetic tape. A paperless format requires a file-naming convention to account for revisions and overall design control. Large holes should be profiled rather than drilled. The designer should check with the fabrication shop to determine the maximum size hole to be drilled.

Fabrication hole sizes for plated versus unplated holes can be automatically determined by a computer program. The designer indicates a plated or unplated hole and the desired finished hole size and the program adds the necessary dimension to obtain plating clearance in the hole. The format for the command structure for drill tapes from the tooling processor should be the same as the CAD system's. Consistency is important for the manufacturing process. Leader feed between turrets, and trailing end will be as follows:

- Leader will contain the part number and revision level from the run card of the digitized data followed by the date and time with only two character spaces between each. After the date and time there will be two spaces, then a number representing the setup along with a letter, i.e., "1A," representing the first tape of the setup. The date and time are critical to ensuring that the most current drill tape is being used in the manufacturing process.

- There should be approximately 10 in (25.4 cm) of tape feed, all tracks punched except track B, between the setup number and the first coded command.

- There should be approximately 10 in (25.4 cm) of tape feed between turret changes.

- At the end of each setup, there should be approximately 10-in of tape feed followed again (for additional setups) by the part number (human readable), revision, and date and time as above.

- Test coupon holes should be representative of the plated through-holes on the printed circuit board and should be the first and last holes drilled in the command string for the turret.

- Each string of commands for each setup will begin with a stop rewind command (tracks 1, 2, and 4) and end with a rewind (M30) command.

- Usually if a setup contains more than 1500 holes, the program will put a stop (M30) on the tape, add 10 in (25.4 cm) of tape feed, increment the letter representing the tape number of the setup, then repeat the first two steps above.

10.13 Profiling

Various methods of profiling the printed circuit board outline are used, such as routing, blanking, and shearing. Control signals for routing should be programmed to make internal cuts first. Routers can be controlled to leave small "breakaway" webs around the perimeter of the printed circuit board or remove all of the material. If manufacturing techniques are such that printed circuit boards are to be assembled while still in the panel, the "breakaway" method is used.

10.14 Test

Database information is also used for the generation of automatic test procedures for the blank printed circuit board. Hole locations and printed circuit board interconnect can be verified on a "bed-of-nails" type fixture or a design dedicated fixture.

10.15 Automatic Insertion Tapes

Postprocessing programs are used to locate and orient components for automatic insertion. The programs will accommodate either single or multiple image assembly techniques. Part sequencing requirements should be considered.

References

1. "Automated Design Guidelines," IPC-D-390, Revision A, February 1988, Institute for Interconnecting and Packaging Electronic Circuits, Lincolnwood, Ill.
2. Robert Dean and Howard Schutzman, "A Layman's View of Placement and Routing Algorithms," Personal CAD Systems, Inc., San Jose, Calif.
3. "Printed Board Description in Digital Form," IPC-D-350, Revision C, November 1987, Institute for Interconnecting and Packaging Electronic Circuits, Lincolnwood, Ill.

Documentation

11.1 Introduction[1]

Whether a printed circuit board is to be a one-of-a-kind prototype or a high-volume production article, it should have some informative type of documentation describing the means to an end. Exactly what should be documented and how it should be prepared must be given serious consideration if a quality product is to be produced within a given budget and schedule. A printed circuit board documentation package should include at least

- Master drawing requirements
- Board definition
- Artwork and phototooling
- Solder mask requirements
- Legend requirements
- Specifications
- Automated techniques (when applicable)

11.2 General Considerations[2]

Too little printed circuit board documentation results in excessive scrappage due to misinterpretation, decreased productivity due to efforts to fill information gaps, and loss of uniform configuration due to "word-of-mouth" manufacturing. Information becomes too dependent on individuals rather than documentation. Too much documentation results in increased drafting hours and labor costs, and decreased manufacturing productivity due to time-consuming interpretation of overemphasized and confusing drawings.

Adequate documentation conveys to the user the basic electromechanical design concept, the type and quantity of parts and materials required, special manufacturing instructions, and up-to-date revisions. The use of adequate documentation offers increased profits by enabling schedule and budget commitments to be met, resulting in satisfied customers who receive quality products on time and at a fair price.

11.2.1   Board types

The detail required in the documentation package varies with the type of printed circuit board being produced. The standard industry types are

- Type 1—Single-sided board (rigid or flexible)
- Type 2—Double-sided board (rigid or flexible)
- Type 3—Multilayer board (rigid or flexible)
- Type 4—Multilayer board (rigid-flexible combination)

11.2.2   Documentation classes

There are three classes of documentation requirements, reflecting progressive increases in sophistication of the drawing package. These classes are minimal documentation, moderate documentation, and full documentation. Selection of class should be based on the minimum need, recognizing that less sophisticated classes require more coordination and communication between user and vendor. Requirements for documentation should be specified in the contract or order used to procure documentation, equipment, or both.

11.2.2.1   Class A—minimal.   In a minimal drawing package, the goal is to prepare the information at the lowest possible drafting cost without compromising the design and integrity of the printed circuit board. Since this type of documentation is primarily used by model shops for quick-reaction prototype development, the drawings should be neat, legible, and informative, but not pretty. (Pretty drawings increase drafting costs and convey no more information.)

This class of documentation usually consists of layout and artwork only. Class A documentation is usually used for internal use and requires a good deal of coordination between the user and manufacturer of the printed circuit board. Information may be incomplete in some instances and relies heavily on in-house manufacturing processes, such as standard material, standard plating processes, and standard tolerances.

Class A documentation is suitable for the applications where the only requirement is that the manufacturer can produce a functional product from information supplied. It may include, as a minimum, the designer's layout with manufacturing notes and instructions and a single-image artwork master.

11.2.2.2 Class B—moderate. A class B documentation package consists of the complete printed circuit board definition without any description of the manufacturing allowances that have been incorporated into the design. Contractual drawing requirements may apply. Quality conformance coupons may be defined by the design; their position in relationship to the printed circuit board or the manufactured panel is optional.

The class B documentation package requires sufficient clarity such that the information may be reviewed by the printed circuit board manufacturer in order to establish product producibility using the artwork or other tooling supplied. Since class B documentation is manufacturer-sensitive, responsibility for various aspects of the manufacturing cycle should be agreed to between user and manufacturer or vendor. Class B documentation is specifically prepared to convey maximum information to the manufacturer; includes a master drawing, all manufacturing notes in a single- or multiple-image artwork master, performance specifications may be referenced, and contractual drawing requirements may apply.

11.2.2.3 Class C—full. Class C is a fully documented reprocurement package to the extent that the information is self-sufficient and may be sent to multiple vendors, with each reproducing the identical product. This documentation package requires that the full manufacturing allowances be disclosed and documented. Quality conformance coupons are mandatory, as required by the design, with the location fixed on the master drawing and artwork establishing the relationship between coupons and the board.

Class C documentation includes a formal master drawing and may include a single- or multiple-image production master, magnetic tape, NC instructions, reference to material requirements, dielectric constant, glass-style resin content, etc., as well as electrical test data, performance testing, and sampling plan callouts. In addition, contractual drawing requirements may apply.

11.2.3 Documentation Categories

Documentation usually relates to two major categories: engineering documentation and manufacturing documentation that may be supplied to the manufacturer as a supplement to the engineering definition. Since the manufacturing documentation is generally tailored to a specific manufacturing facility, process, or equipment, it is not normally part of a reprocurement package (i.e., technical data package).

11.2.3.1 Engineering documentation. The basic engineering documentation package usually includes:

- Master drawing and the documentation invoked thereon, such as
    - *a.* Material specification
    - *b.* Process specification
    - *c.* Performance specification
    - *d.* Test methods/specification
    - *e.* Part documentation
- Single- or multiple-image artwork

11.2.3.2 Manufacturing documentation. The basic manufacturing documentation package usually includes:

- Production master
- Soft tooling (drill tapes)
- Manufacturing layout instructions

11.3 Master Drawing

The master drawing should include, as appropriate, the information described in Table 11.1 either by direct delineation or by reference to other documents which become a part of the engineering documentation set. One part number should be assigned to the master drawing which, basically, describes the end product(s). When the drawing depicts multiple versions of a printed circuit board, a unique part number (usually the basic drawing number supplemented by dash numbers) should be used to identify each version. References may be established, where applicable, depending on classification of documentation package to performance specifications, end-product requirement specifications, engineering layout with manufacturing instructions, or quality conformance test coupons.

11.3.1 Datums

There should be a minimum of two datum features to establish the mutually perpendicular datum reference frame for each printed circuit board. Critical design features may require the use of more than one set of datum references. The master drawing should establish the relationship and acceptable tolerance between all datum and circuit features. All datum features should be located on grid or establish

TABLE 11.1   Master Drawing Detail Descriptions

Characteristic	Requirement
A. Board details	1. The type, size and shape of the printed circuit board and all tolerance conditions thereof. 2. Dielectric separation between layers. 3. Bow and twist allowances. 4. Overall board thickness requirements including tolerances.
B. Materials	1. Type class and grade of material, including color if appropriate. 2. Plating and coating material(s) type and thickness(es). 3. Marking inks or paint and permanency.
C. Hole details	1. The size, location and tolerances for all holes. 2. The plating requirements for all holes. 3. The hole cleaning requirements including if etchback is required or permitted.
D. Conductor definition	1. Shape and arrangement of both conductor and non-conductor patterns on each layer of the printed board. (Copies of the production master[s] or copies of artwork may be used to define these patterns) 2. Separate views of each conductor layer. 3. Dimensions either specifically or by reference to a grid system, for any and all pattern features not controlled by hole size and location. 4. Minimum conductor width and spacing on the finished printed circuit board. 5. Dimensions and tolerances for critical pattern features which may affect circuit performance because of distributive inductance or capacitance effects. 6. Conductor layer identification starting with the primary side as Layer 1. If there are no conductors or lands on the primary side, the next conductive layer shall be Layer 1. (For assemblies with components on both sides the most complex or densely populated side shall be Layer 1.)
E. Marking	1. Printed board identification marking. 2. Size, shape and location of reference designation and legend marking, if required. 3. Serialization and revision level marking (when required).
F. Processing conditions	1. Processing allowances that were used in the design of the printed circuit board including but not limited to: *a.* Conductor width allowance *b.* Conductor spacing allowance *c.* Land/hole fabrication allowance *d.* Solder mask or cover layer registration allowance. 2. Applicable processing-specifications. 3. Location of quality conformance coupon or circuitry
G. Design concepts	1. Maximum rated voltage (maximum voltage between two non-connected adjacent conductors with the greatest potential difference). 2. Identification of testing and test point location. 3. Modular grid system(s) used to identify all holes, test points, lands and overall board dimensions. *a.* Metric grid—2.0, 1.0, 0.5 or multiples of 0.1 mm *b.* Inch based grid—0.100, 0.050, 0.025 or multiples of 0.005 inches (see Section H-3 of this table.
H. Documentation practices	1. Terms used on the master drawing shall conform to this standard, IPC-T-50 or ANSI-Y 14.5. 2. Notes either included on the first sheet(s) or location of notes specified on the first sheet(s). 3. Indication of grid using *X-Y* control dimensions or grid scales. To maintain clarity of conductor patterns shown on the master drawing, overall grids shall not be reproducible.

SOURCE: IPC-D-325.

grid criteria, as defined on the master drawing, and should be within the outline of the printed circuit board.

11.3.2 Processing allowances

For class C documentation, the processing allowances that were considered in the design and artwork preparation for the printed circuit board should be documented and defined on the master drawing in either note form or by reference to another drawing which contains artwork specifications. This information should be expressed in terms of the maximum variation between the end-product conductor widths and spacings and what may appear on the artwork; the minimum land, in reference to the drilled or plated hole and what may appear on the artwork; or any other feature conditions considered in the design where the variation between end-product and artwork configuration play a role in the producibility of the printed circuit board.

- *Conductor width allowance:* List the smallest allowance for the particular design.

- *Conductor spacing allowance:* List the largest negative number or, if all numbers are positive, list the smallest positive number.

- *Land fabrication allowance:* List the smallest number resulting from the calculation.

- *Hole size tolerance:* List the smallest difference.

- *Aspect ratio:* List the largest number resulting from the calculation. Carry to one decimal place.

- *Maximum rated voltage:* List the largest number resulting from the calculation.

11.3.3 Quality conformance test circuitry

Quality conformance test circuitry should be included on the master drawing and artwork for all class C documentation and may be included for class B. The test circuitry for the printed circuit board should reflect the design of the board and all manufacturing processes, such as drilling, plating, etching, fusing, ground/voltage, planes, separately fabricated layers, and permanent coatings. The specific quality conformance test circuitry should be dependent on the type of product being designed and the particular design specification invoked.

11.4 Board Outline Drawing[2]

The printed circuit board outline drawing is used to convey information about the shape and drill pattern of the board. All outside shapes with dimensions, all tolerances, and all common tooling features are dimensioned. Each hole should be shown with a code describing the hole size. The board outline should contain at least two special datum holes that are represented by an easy-to-center target such as a printer's bullseye or quadrant. One of these datums should be (0.0) and all holes and features referenced to it. The datums should appear on all artwork layers.

Part of the board outline may also include the X and Y coordinates for all drilled holes to allow drill tape checking. The board outline drawing may also make reference to surface finish, thickness requirements, solder mask, laminate, and related part drawings such as heat sinks, coatings, and assemblies.

11.4.1   Profile dimensions

The depictions of the magnitude of feature size, positions, and form, and the tolerances associated with these attributes, should show:

- *Thickness:*   The thickness for type 1 and type 2 boards may be given on an edge view. The thickness for type 3 or 4 boards may either be presented in an edge view or it may appear on the construction view (Fig. 11.1). Dimensions should clearly indicate where they apply, i.e., overplating, minus plating, etc.

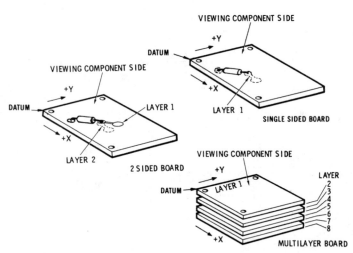

Note: For those printed board assemblies that have components on both sides, any side may be designated as the primary side and is used to define layer 1.

Figure 11.1   Printed circuit board viewing.   (*From IPC-D-325.*)

- *Shape:*   The board edges should be shown in accordance with IPC-D-300 or other appropriate end-product requirements.

11.4.2 Dielectric spacing

The master drawing should address the dielectric spacing between adjacent layers to the detail necessary for the correct electrical functioning of the printed circuit board, particularly the layer-to-layer dielectric spacing represented by prepreg material.

11.4.3 Construction

If the regulatory design specification on the board does not establish the desired constraints on the following items, each multilayer board drawing should contain information that:

- Identifies the number and location of each layer (Fig. 11.1).

- Establishes the location of all materials (Fig. 11.2). If this view contains end-product thickness requirements it should clearly indicate where and when this dimension applies, i.e., "overplating," "prior to fusing," or "excluding surface conductors."

- Describes material thickness for each layer (Fig. 11.3).

11.4.4 Board materials.   The documentation should specify the applicable materials in sufficient detail that the end product will meet the desired performance requirements and preclude the selection of un-

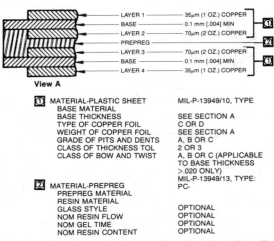

Figure 11.2 Printed circuit board material callout. (*From IPC-D-325.*)

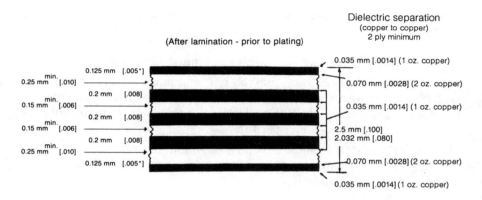

Figure 11.3 Printed circuit board dielectric thickness dimensioning. (*From IPC-D-325.*)

desired material by the printed circuit board fabricator. This may be done on the master drawing or a separate material specification should be developed which is called for by the master drawing. Documentation should not call out material attributes that vary with the printed circuit board processing facility and will not affect the end-product performance. Shown below are some typical examples of material callouts on the master drawing to describe the material.

- Material[3] (plastic sheet)—per MIL-P-13949/4.
- Composite board material[4] (CM-3)—0.060 in-thick per IPC-L-112.
- Metal clad Dielectric[5]—0.001 in thick with 1-oz copper on each side, per IPC-FC-241.

In some instances, for critical printed circuit board performance, it may be required that the documentation provide the definition of the specific resin, glass style, adhesive, or other attributes necessary to ensure the dimensional characteristics and physical or electrical performance of the end product. Caution should be used when defining these attributes on the master drawing in that their use requires a complete understanding of the manufacturing process and the physical parameters of printed circuit board fabrication.

11.4.4.1 Copper-clad laminate material. The detail required to adequately define the copper-clad laminate material will vary with the specific application. Material callouts should be made considering the following:

- *Base material type (GF, GI, FR, etc.):*   must always be given.
- *Base material thickness:*   may be left as a manufacturer's option if the end-product thickness requirements are well-defined on the master drawing and if the electrical performance requirements (voltage, impedance, etc.) permit.
- *Grade of pits and dents:*   usually a manufacturer's option, especially if end-product conductor defects are separately defined in a procurement specification.
- *Class of thickness tolerance:*   manufacturer's option, unless required for electrical performance.
- *Class of bow and twist:*   manufacturer's option, to meet the bow and twist requirements of the end-product printed circuit board.

11.4.4.2 Prepreg material. Callouts for prepreg material should be made considering the following:

- *Prepreg material type:*   must always be given on the master drawing.
- *Prepreg glass style:*   usually a manufacturer's option.
- *Prepreg nominal resin flow:*   usually a manufacturer's option.
- *Prepreg nominal gel time:*   usually a manufacturer's option.
- *Prepreg nominal resin content:*   usually a manufacturer's option.

11.4.4.3 Metal-clad flexible dielectrics[5]. Unless otherwise indicated, metal-clad flexible dielectrics should conform to IPC-FC-241. A complete composite designation of the metal-clad flexible dielectric defined in IPC-FC-241 should be required on the master drawing.

11.4.4.4 Cover layers for flexible printed wiring[6]. Unless otherwise indicated, the cover layers for flexible printed wiring should conform to IPC-FC-232. The complete composition designation of the flexible cover layer defined in IPC-FC-232 should be required on the master drawings.

11.4.4.5 Flexible adhesive bonding films[7]. Unless otherwise indicated, the flexible adhesive bonding films should conform to IPC-FC-233.

The complete designation of the flexible adhesive bonding film defined in IPC-FC-233 should be required on the master drawing.

11.4.4.6 Unclad laminates[8]. Unless otherwise indicated, unclad laminate material intended for use in additive manufacturing processes should conform to IPC-AM-361. The complete designation for unclad laminates should be required on the master drawing.

11.4.4.7 Copper foil[9]. Unless otherwise indicated, when copper foil is specified on the master drawing, the requirements of IPC-CF-150 should be invoked. The appropriate class for copper foil defined in IPC-CF-150 should be designated on the master drawing: classes 1 through 4 are usually used for rigid materials, and classes 5 through 8 are usually used for flexible materials.

11.4.4.8 Copper film[10]. For additive circuitry, copper film should meet the requirements of IPC-AM-372. The appropriate copper film requirements should be defined on the master drawing.

11.4.4.9 Solder mask[11]. If solder mask is required on the printed circuit board, drawing notes are required to cover the solder mask material. The mask material should be per IPC-SM-840. Type may be left optional. Class should be specified, for example, "Solder mask on side indicated per IPC-SM-840. Type 1 or 2, class 3."

11.4.5 Holes

All holes should be identified on the master drawing so that each hole is identified and size, location, and tolerance are clearly prescribed.

11.4.5.1 Hole size. The finished size of all holes should be prescribed. Specified hole dimensions should apply after all required platings and treatments, such as solder fusion. Plated, unplated, or optional holes should clearly be identified on the master drawing. If holes do not go through the entire board thickness, such as buried vias, the master drawing should clearly define the layers through which the holes must be fabricated. A method of expressing the hole size tolerance is to give both the upper and lower size limits. (See IPC-D-300 for examples of size ranges and tolerance application.[12])

11.4.5.2 Hole location. All holes should be located with respect to single or secondary grid systems in accordance with the appropriate design standard. Holes shown to be centered on lands and which lack expressly dimensioned locations are understood to be ideally located on the same grid-line intercepts that best describe the ideal locations of the lands. All holes not located on grid or centered on lands should be expressly dimensioned on the master drawing, with location tied to the coordinate system established by the printed circuit board datums.

Note. A hole master phototool is recommended whenever a design is likely to be produced by a fabricator who might lack the ability to use the NC drill tape produced by the designing activity.

Location tolerance is preferably expressed using positional tolerances. The tolerance on hole location is preferably shown on the master drawing hole chart.

Note. When the minimum annular ring is the only pertinent constraint on hole location, the tolerance need not be stated.

For some Class C documentation packages, complete X-Y description may be required of the printed circuit board. In this instance a table such as shown in Fig. 11.4 should be prepared and correlated to the hole chart describing the various hole sizes.

11.4.5.3 Etchback. Any type 3 or 4 board that specifically requires or permits an etchback operation, as opposed to normal hole-cleaning requirements, must contain a drawing note invoking this requirement and specifying the minimum and maximum etchback dimensions.

11.4.6 Circuitry

Circuit patterns may be documented using any of several styles for artwork description. The master drawing should contain or reference separate views of each conductor layer, clearly showing the shape and arrangement of both conductor and nonconductor patterns or elements. Any and all pattern features not controlled by the hole sizes and locations should be adequately dimensioned, either specifically or by notes.

11.4.6.1 Minimum conductor width. Master drawings for printed circuit boards should contain a note indicating the minimum acceptable conductor width on the end product.

11.4.6.2 Minimum spacing between conductors. Master drawings should contain a note indicating the minimum acceptable spacing between conductors on the end product.

FEATURE	LOCATION		FEATURE	LOCATION	
	X	Y		X	Y
PATTERN G			PATTERN A		
A1	1.000 (25.40)	3.000 (76.20)	1	5.600 (142.24)	3.100 (78.74)
A2	1.125 (28.57)	3.000 (76.20)	2	5.400 (137.16)	3.100 (78.74)
A3	1.250 (31.75)	3.000 (76.20)	3	5.400 (137.16)	2.750 (69.85)
A4	1.375 (34.93)	3.000 (76.20)	4	5.400 (137.16)	2.400 (60.96)
B1	1.000 (25.40)	2.625 (66.68)	5	5.600 (142.24)	2.400 (60.96)
B2	1.125 (28.57)	2.625 (66.68)	6	5.900 (149.86)	2.400 (60.96)
B3	1.250 (31.75)	2.625 (66.68)	7	6.100 (154.94)	2.400 (60.96)
B4	1.375 (34.93)	2.625 (66.68)	8	6.100 (154.94)	2.750 (69.85)
B5	1.500 (38.10)	2.625 (66.68)	9	6.100 (154.94)	3.100 (78.74)
B6	1.625 (41.28)	2.625 (66.68)	10	5.900 (149.86)	3.100 (78.74)
B7	1.750 (44.45)	2.625 (66.68)			
B8	1.875 (47.63)	2.625 (66.68)	PATTERN H		
B9	2.000 (50.80)	2.625 (66.68)			
B10	2.125 (53.98)	2.625 (66.68)	1	1.160 (29.46)	6.400 (162.56)
B11	2.250 (57.15)	2.625 (66.68)	1'	1.160 (29.46)	5.000 (127.00)
B12	2.375 (60.33)	2.625 (66.68)	2	1.825 (46.36)	6.300 (160.02)
B13	2.500 (63.50)	2.625 (66.68)	2'	1.825 (46.36)	5.800 (147.32)
C1	1.000 (25.40)	2.250 (57.15)	3	1.675 (42.55)	6.300 (160.02)
C7	1.750 (44.45)	2.250 (57.15)	3'	1.675 (42.55)	5.800 (147.32)
C13	2.500 (63.50)	2.250 (57.15)	4	2.850 (72.39)	3.700 (93.98)
D1	1.000 (25.40)	1.875 (47.63)	4'	2.850 (72.39)	4.200 (106.68)
D13	2.500 (63.50)	1.875 (47.63)	5	1.675 (42.55)	3.700 (93.98)
E1	1.000 (25.40)	1.500 (38.10)	5'	1.675 (42.55)	4.200 (106.68)
E13	2.500 (63.50)	1.500 (38.10)	6	1.825 (46.36)	3.700 (93.98)
F1	1.000 (25.40)	1.125 (28.57)	6'	1.825 (46.36)	4.200 (106.68)
F2	1.125 (28.57)	1.125 (28.57)	7	0.650 (16.51)	6.300 (160.02)
F3	1.250 (31.75)	1.125 (28.57)	7'	0.650 (16.51)	5.800 (147.32)
F4	1.375 (34.93)	1.125 (28.57)	8	2.850 (72.39)	6.300 (160.02)
F5	1.500 (38.10)	1.125 (28.57)	8'	2.850 (72.39)	5.800 (147.32)
F6	1.625 (41.28)	1.125 (28.57)	9	0.650 (16.51)	3.700 (93.98)
F7	1.750 (44.45)	1.125 (28.57)	9'	0.650 (16.51)	4.200 (106.68)
F8	1.875 (47.63)	1.125 (28.57)	10	2.340 (59.44)	3.600 (91.44)
F9	2.000 (50.80)	1.125 (28.57)	10'	2.340 (59.44)	5.000 (127.00)
F10	2.125 (53.98)	1.125 (28.57)			
F11	2.250 (57.15)	1.125 (28.57)	PATTERN B		
F12	2.375 (60.33)	1.125 (28.57)			
F13	2.500 (63.50)	1.125 (28.57)	1	3.500 (88.90)	5.625 (142.87)
G10	2.125 (53.98)	0.750 (19.05)	2	3.625 (92.08)	5.625 (142.87)
G11	2.250 (57.15)	0.750 (19.05)	3	3.750 (95.25)	5.625 (142.87)
G12	2.375 (60.33)	0.750 (19.05)	4	3.875 (98.43)	5.625 (142.87)
G13	2.500 (63.50)	0.750 (19.05)	5	4.000 (101.60)	5.625 (142.87)
PATTERN F			6	3.625 (92.08)	5.875 (149.22)
1	5.500 (139.70)	0.750 (19.05)	7	3.750 (95.25)	5.875 (149.22)
2	5.500 (139.70)	1.000 (25.40)	8	3.875 (98.43)	5.875 (149.22)
3	5.500 (139.70)	1.250 (31.75)	9	4.000 (101.60)	5.875 (149.22)
4	5.500 (139.70)	1.500 (38.10)	10	3.500 (88.90)	6.125 (155.57)
5	5.500 (139.70)	1.750 (44.45)	11	3.625 (92.08)	6.125 (155.57)
6	6.000 (152.40)	0.750 (19.05)	12	3.750 (95.25)	6.125 (155.57)
7	6.000 (152.40)	1.000 (25.40)	13	3.875 (98.43)	6.125 (155.57)
8	6.000 (152.40)	1.250 (31.75)	14	4.000 (101.60)	6.125 (155.57)
9	6.000 (152.40)	1.500 (38.10)			
10	6.000 (152.40)	1.750 (44.45)			

Data in parenthesis ( ) is expressed in millimeters,
reference information is in brackets [ ].

ALL DIMENSIONS ARE BASIC.

Figure 11.4 Example of printed circuit board *X-Y* coordinate feature location chart. (*From IPC-D-325.*)

11.5 CAD Documentation[13]

The conversion program can automatically insert drawing features such as standard notes into the finished drawing. Magnetic tapes used to plot the general notes, board outline, board trim dimensions, and other peripheral details should be sent to the plotter prior to the printed circuit board plot tape. A master of these can be plotted on the required format and sufficient copies printed for use when the printed circuit board tapes are finally received. Tailored notes or images can follow in the standard format, while the board description data provides the special addition for notes or figures.

A code name can identify each repeated format. When the notes and drawing details are plotted, this code will be marked in the border of the format by the plotter operator.

11.5.1 Schematics

Schematic output from CAD systems can vary with the datum. In order to maintain consistent ink-plotting procedures, the plotter starting point (origin) for all schematics and formats should be at the border in the lower left corner of the drawing form. Also, the scale of the schematic output directly from the CAD system may not be the scale desired for ink-plotting requirements. The schematic may scale to a system default. To prevent accidental scaling that would expand the drawing beyond plotter limits, check to assure that each assembly and drill plot will fit the plotter after scaling. Some CAD systems have check programs to do this. If the CAD system does not, a custom program can be written, or the offsets can be calculated manually. If this check is not done, the plotter operator will waste valuable plotting time trying to correct the overscale.

11.5.2 Detail drawings

The design database can generate detailed dimensional drawings of the printed circuit wiring board. Drill views should be drawn looking at the component or primary side of board detail. When components appear on both sides of the board, the primary side is the one with the most number of active components.

11.5.2.1 Assembly drawings.

Assembly views should be drawn looking at the component side. When components appear on both sides of the board, the primary side is the one with the highest number of active components.

11.5.2.2 Solderless (wire) wrap interconnect lists.

Programs automatically determine wirewrap interconnects including routing and wire lengths from a basic printed circuit wiring board database that indicates the positional location of components. Interconnect of the components (printed circuit conductors) is not necessary; however, a net list of connections must be in the database.

11.5.3 Final checking

In addition to all previous checking, output devices that take data from the system and execute the final step of a portion of the manufacturing cycle require specific checks. Items in this category are referred to as output and include, but are not limited to

- Artwork photoplot
- Shape-fill—shapes defined by lines and arcs
- Aperture selection
- Conductor tracks
- Solder mask
- Silk screen
- CNC/NC drill (tape or drill mask)
- CNC/NC insertion tape
- Documentation plots—schematic for technical publications and assembly drawings
- Generate bill of materials (BOM) listing—this requires a general routine to customize BOM listing for each user.
- Generate manufacturing test patterns—test-pattern-generating routines must be customized for each type of test machine.

Checking this type of data requires complete understanding of the rules and commands used by a particular piece of equipment. Not only does the CNC/NC equipment adhere to a certain command structure, many companies adopt standard conventions related to a particular task.

11.6 Data Management[13]

Database consistency between the logical database (schematic), the physical database, the film masters, and the physical board is critical for successful CAD implementation. One benefit of CAD systems is ease in design revision. This benefit is negated if the printed circuit board has a corrupt database. Computer programs for a completely automated on-line, interactive database system provide a highly technological organization with a tool for managing the design configuration of end items and/or products. Such a system can baseline both the documents and parts that define a proposal bid and continue to track the design through engineering development and production phases.

A virtually unlimited number of configuration items can be maintained. The system can guide the designer in the selection of parts, materials, and processes that the company and/or the customer prefers to use. The system can also provide other data such as specifications, reference documents, and standard notes required for a complete parts list. Such a system can be expanded to assign both document and part numbers and track documents through design, release, reproduction, and storage. It can incorporate changes into the master document in a timely manner to meet any company or contractual requirements. This can be a front-end system in a complete information management system operation and can provide automated interfaces and data transfer with the manufacturing bill of materials, configuration and data management trace systems, estimated inventory and order management systems, logistic provisioning and logistic support analysis systems. It can also interface with other CAE/CAM systems.

A document file is a computer data file that contains status information on all hard copy and computer data file documents. The document file contains related data for hard copy on document revision level, document size, total number of sheets, document type, security class, document location, work authorization, and change history.

Information status of computer files is similar and includes file revision level, file size, file type (if system has identification for different files), security class (file location, alternate disk, magnetic tape, other computers, etc.). File access and change history require authorization. Maintenance of this file originates from the drawing/part/change number assignment process and the drawing/part/change release process.

11.7 Assembly Drawings[2]

The assembly drawing is usually not sent to printed circuit board fabricators unless they will also do the stuffing and soldering. The assembly drawing shows all parts identified by their reference designators (C1, U48, etc.) in position. The assembly drawing should show the datum positions, and if automatic insertion is anticipated, should show X and Y coordinates for some part of each component.

11.8 Processing Parameters[1]

Requirements of the end product printed circuit board, related to processing parameters, should be documented on the master drawing or a separate specification should be developed to provide the general information on process-related requirements.

11.8.1   Plating

The master drawing should contain requirements pertaining to the type and thickness of all plating layers, including detailed specification requirements if applicable. Plating requirements should be as determined by the appropriate design standard. For classes B and C type documentation, minimum platings should be described on both the surface and in the hole. This consists of initial basic copper plating, as well as any overplate used in the manufacturing.

11.8.2   Application of solder mask

The solder mask characteristics related to the process may be detailed on the master drawing or a separate specification may be developed.

11.8.3   Legend

The master drawing should include the specification requirement for all supplemental marking, that is, markings other than those etched in the pattern or a separate process-oriented specification should be developed. The information shall include the kind of ink or paint, the color, and the side(s) of the printed circuit board to be marked. If the marking is to be made by screen printing or similar method a layout of the marking should be included as the last sheet(s) of the artwork master and should be referenced in the marking note.

The master drawing should specifically indicate the location for the manufacturer's code and all other required traceability markings on the printed circuit board and on the test coupon, if applicable. The part number change letter and its location should be indicated on the drawing. A drawing note should indicate the acceptable methods of marking (ink, rubber stamp, electric pencil, etching, etc.) and color if applicable. Traceability marking should include a lot number and/or some method of serializing traceability.

11.9   Performance Specifications

The master drawing should reference the performance specifications necessary for describing the parameters of the printed circuit board. In some instances, industry and military specifications are preferred as they represent a consensus. In-hour or other documents may also be developed for those requirements which are outside the industry-agreed-to parameters. Some of the types of performance and end-product requirements necessary to be defined in the process specification include

- Visual examinations
- Workmanship
- Solderability
- Thermal stress
- Testing
- Dimensional requirements
  - *a.* Hole pattern
  - *b.* Bow and twist
  - *c.* Conductor spacing
  - *d.* Conductor pattern
  - *e.* Layer-to-layer registration
  - *f.* External annular ring (unsupported hole or plated through hole)
  - *g.* Solder mask thickness
  - *h.* Tin-lead or solder coating thickness
- Physical requirements
  - *a.* Plating adhesion
  - *b.* Conductor edge overhand
  - *c.* Bond strength
  - *d.* Microsectioning construction integrity (plated through-hole evaluation, plated copper thickness, layer-to-layer registration, and plating voids)
  - *e.* Conductor thickness
  - *f.* Resin smear and etchback
  - *g.* Undercut
  - *h.* Internal annular ring
  - *i.* Dielectric layer thickness
  - *j.* Laminate voids
  - *k.* Resin recession
- Electrical and environmental requirements
  - *a.* Moisture and insulation resistance
  - *b.* Dielectric withstand voltage
  - *c.* Circuit continuity
  - *d.* Insulation resistance (circuit shorts)
  - *e.* Cleanliness
  - *f.* Ionic/organic contamination
- Repair

Figure 11.5 incorporates the inspection requirements of the industry document only and provides additional requirements. Some companies may develop in-house specifications which will be used in the manufacturing processes. This technique may be helpful for captive manufacturing but may also be used for external manufacturing. Figure 11.6 shows an example of a multilayer in-house specification.

NOTES:

1. MATERIAL:
   EPOXY GLASS LAMINATE TYPE NEMA G10FR .062 THK.
   1 OZ. COPPER CLAD ON 2 SIDES

2. DRILLING:
   SEE HOLE LEGEND.
   ALL HOLE SIZES ARE FINISHED, PLATED-THROUGH SIZE.

3. PLATING:
   THE WALL OF THE PLATED THROUGH HOLES SHALL BE COPPER, THE THICKNESS OF WHICH SHALL NOT BE LESS THAN .001 INCH.

   ALL CONDUCTOR PATTERNS, EXCLUSIVE OF CONNECTOR FINGERS, SHALL BE COVERED BY ELECTRO-DEPOSITED SOLDER WHOSE COMPOSITION IS 70% TIN MAXIMUM AND 30% LEAD MINIMUM. THE APPEARANCE TO BE BRIGHT REFLECTIVE WITH A THICKNESS OF .0003 INCH MIN. THE SOLDER SHALL BE REFLOWED.

   CONNECTOR FINGERS SHALL BE .000020 MIN. GOLD OVER .000150 MIN. NICKEL.

4. MARKING:
   COMPONENT . OUTLINE AND REFERENCE DESIGNATIONS SHALL BE SCREENED IN WHITE EPOXY INK ON THE COMPONENT SIDE OF THE P.C. BOARD.

5. FINAL FABRICATION:
   FABRICATE TO THE DIMENSIONS SHOWN. CARD EDGES SHALL BE SMOOTH AND FREE OF BURRS.

   BEVEL CONNECTOR CONTACT EDGE .015 X 45° BOTH SIDES AND ENDS. A PROTECTIVE COATING OR OTHER PROTECTIVE MEANS SHALL BE PROVIDED TO INSURE GOOD SOLDERABILITY AFTER STORAGE.

   SOLDER MASK MUST NOT FLAKE OR SEPARATE DURING SOLDERING. TYPICAL SOLDER BATH: 480°F (248°C) FOR 1 SECOND DURATION.

8. INSPECTION REQUIREMENTS:
   REFER TO IPC-SD-320.

Figure 11.5   Example of double-sided printed circuit board drawing notes.   (*From IPC-D-325.*)

11.10   Test Methods

Testing procedures may be included as a part of the master drawing or a separate procedure may be developed which becomes part of the engineering documentation. IPC-TM-650 is the IPC test methods manual and may be used for providing testing requirement information to the manufacturer.

1. GENERAL

A. This board ☐ has ☐ has not been designed for automatic insertion of components.

B. Unless otherwise specified, the finished board shall meet:

(1) Master 1X hole symbology plot.

(2) Specific requirements of other sheets of this board drawing package.

(3) In general, the latest revision of applicable standards of the Institute for Interconnecting and Packaging Electronic Circuits (IPC). The tables referenced below are from IPC-D-300G.

C. A NC drill tape defining hole diameter size(s) and locations ☐ has ☐ has not been supplied.

(1) The format of the tape is _____ .

(2) The information on the tape ☐ has ☐ has not been sorted by hole diameter size.

(3) The information on the tape ☐ has ☐ has not been optimized for minimizing drill movement.

2. PHOTOGRAPHIC REDUCTION

A. Only the 1X master patterns supplied are to be used for photographic reproduction of the circuit pattern(s), hole symbology plot and silk screen legend. No changes are to be made to the 1X master patterns without prior approval. The substitution of prints or other copies of the 1X master patterns is not acceptable.

B. The 1X master conductor patterns are to be supplied to the vendor on ☐ mil thick Estar based film with a stabilized gelatin emulsion ☐ glass plate.

3. CONSTRUCTION

A. The following table defines the multilayer "sandwich" of the completed printer wiring board. The artwork identification refers to the name of the layer as it appears on the master artwork pattern supplied. Layer 1 is always the component side. The last layer is always the non-component side.

LAYER NUMBER	ARTWORK IDENTIFICATION	NOMINAL FOIL WEIGHT (OZ/FT2)	FINAL THICKNESS BETWEEN LAYERS
1	Component Side		
2			±
3			±
4			±
5			±
6			±
7			±
8			±
9			±
10			±
11			±
12			±
13			±
14			

B. The total thickness of the multilayer printed wiring board is _____ .

C. The maximum thickness of the board (including plating) shall not exceed grade _____ .

D. The identification "LAYER _," or "■■■_____" is included on each master artwork pattern within the printed wiring board circuit area. This is a precisely located and configured identification that is used to assure that the layers have been laminated in the correct order and orientation. The bars ([ ]) serve to block out the numbers if the layers are incorrectly positioned. When correctly superimposed on one another after the laminating process, "LAYER 1 2 3 4....." should be readable from the component side.

4. MATERIAL

A. The composite sheet, base material (laminate) and foil shall meet the chemical and dimensional requirements listed below of the latest revision of Military Specification MIL-P-13949F, "Plastic sheet, laminated, metal-clad (for printed wiring boards), general specifications for". **Certification is not required.**

(1) Base material _____

(2) Base Color _____

(3) Grade of Pits and Dents _____

(4) Class of Thickness Tolerance _____

(5) Class of Bow and Twist _____

B. The laminate copper foil weight for the external (component and non-component layers) is a recommendation only. It is the vendor's responsibility to select the correct laminate foil thickness to meet the requirements of this specification (such as final copper thickness, hole plating, conductor width tolerance, etc.).

C. The resin preimpregnated glass cloth (B-stage, pre-preg) shall meet the chemical and dimensional requirements listed below of the latest revision of Military Specification MIL-P-13949F, "Plastic sheet, laminated, metal clad (for printed wiring boards), general specifications for". **Certification is not required.**

(1) Prepreg Material ___PC___

(2) Resin Material _____

5. INNER LAYER SURFACE TREATMENT

A. A heavy oxide (i.e. black oxide) surface treatment is required to increase copper surface area and promote lamination bond strength.

6. CONDUCTORS

A. The narrowest copper conductor width and spacing on the external layers artwork for this board is:

(1) Width: _____ nominal

(2) Spacing: _____ nominal

B. Finished copper conductor width variation on the external layers shall not exceed 25 percent of IX master pattern supplied.

For boards that have fine line conductors (_____ inch wide) on the external layers, the finished width shall be _____ inch minimum.

C. The narrowest copper conductor width and spacing on the internal signal layer(s) artwork for this board is:

(1) Width: _____ nominal

(2) Spacing: _____ nominal

D. Finish copper conductor width variation on the internal signal layers shall not exceed 25 percent of IX master pattern supplied.

E. No defect shall reduce the cross section of a copper conductor by more than 25 percent.

F. No damaged or broken copper conductor may be repaired without prior approval.

7. HOLE AND CONDUCTOR PATTERN REGISTRATION

A. Hole centerline tolerances are grade _____ .

B. The targets around the datum holes on the non-component side shall be within _____ of true position.

C. The layer-to-layer registration between any two layers shall be within _____

D. The minimum annular ring on all land areas on the external layers shall not be less than grade _____ . The thickness of the metal in the plated-thru-hole is included.

E. The minimum annular ring on all land areas on the internal layers shall not be less than grade _____ . The thickness of the metal in the plated-thru-hole is included.

F. The unplated-thru-hole (nonsupported hole) diameter tolerances are grade _____ except where noted.

G. Unless otherwise noted, all holes are designed on a

☐ 2.5mm [0.100 inch] grid   ☐ 1.25mm [0.050 inch] grid
☐ .025 inch grid   ☐ 0.5mm grid   ☐ no grid system.

Figure 11.6 Example of general specification for multilayer printed circuit boards. (*From IPC-D-325.*)

11.11   Parts Documentation

The master drawing may also include supplementary steps to fabricate a finished printed circuit board, such as the addition of heat sinks, terminals, eyelets, stiffeners, and/or miscellaneous hardware. If this option is exercised, the master drawing should show the location

8. SMEAR REMOVAL (Hole Cleaning)

A. The holes in the board must be mechanically or chemically cleaned for the lateral removal of material from the internal layer(s) prior to plating.

B. After plating, there must be no evidence of resin smear or epoxy residue.

C. After plating, there must be a direct bond of the plated copper to the foil copper of the internal layer. A line of demarcation is acceptable.

D. Etchback (removal of epoxy smear *and* glass fibers) is not required but acceptable.

9. PLATING (External Layers Only) ☐ Required ☐ Not Required

A. Plated-thru-holes

(1) Material ☐ Copper ☐ Other _____ .

(2) Thickness (on walls) ☐ 0.025 mm [0.001 inch] ☐ 0.038 mm [0.0015 inch] ☐ Other _____ minimum.

(3) The indicated hole will be used for cross sectioning purposes. A cross sectioned sample ☐ is ☐ is not required to be shipped with each lot.

(4) All holes must be within the size tolerance after all plating and reflow operations have been completed.

(5) Diameter tolerances are grade _____ per Table 8 except where noted.

B. Conductor paths other than printed edge board contacts (fingers):

(1) Copper: final copper thickness (base foil plus plating to be):

☐ 0.043 mm [0.0017 inch] minimum/side

☐ 0.061 mm [0.0024 inch] minimum/side

☐ Other _____

(2) Additional plating:

a. Material:

☐ Fused solder plating (tin lead) _____ percent tin.

☐ Other _____

☐ Other _____

b. Thickness: ☐ 0.008 mm [0.0003 inch] minimum at crown after reflow

c. Contact finger plating:

(1) Material:

☐ Nickel/gold 0.0025 mm [0.0001 inch] minimum, low stress nickel under 0.00127 mm [0.00005 inch] minimum 99.7% pure gold having a hardness of Knopp 140-220).

☐ Other _____

Plating length from leading edge of fingers:

☐ 7.62 mm [0.300 inch] minimum.

☐ Other _____

(3) The leading edges of the contact fingers are to be beveled 0.50 ± 0.25 mm [0.020 ± 0.010 inch] by 45 degrees on both sides of the board.

10. SOLDER MASK ☐ Required ☐ Not Required

A. Solder mask on ☐ non-component side ☐ component side.

B. The solder mask 1X master pattern is supplied and must be used.

C. See solder mask drawing, sheet _____ of _____

D. Mask with (either):

☐ Dry Film (Solvent) _____

☐ Dry Film (Aqueous) _____

☐ Other _____

☐ Other _____

E. The mask shall conform to the board's pad area size and configuration and shall not extend onto the pad area. The mask ☐ shall ☐ shall not extend onto the pad area of via holes.

F. Solder mask material is to be applied as specified by the material.

G. All solder plated boards are to be reflowed prior to application of solder mask.

H. Damaged or missing solder mask may not be repaired without prior approval.

11. SILK SCREEN LEGENDS ☐ Required ☐ Not Required

A. Silk screen legends on ☐ non-component side ☐ component side.

B. The silk screen legend 1X master pattern is supplied and must be used.

C. See silk screen legend drawing, sheet _____ of _____. Screen with _____ screening ink. Material Specification Number _____ (_____).

D. Silk screen must be applied after application of solder mask (if required).

E. The silk screen ink **must not** extend into any plated-thru-hole.

12. SOLDERABILITY

Test per IPC-S-804.

13. CRACKS, CHIPS AND BURRS

A. No cracks are allowed.

B. No chips, cracks or haloing about the periphery of the board shall extend onto the top or bottom face area of the board further than 0.762 mm [0.030 inch] from the board edge.

C. The board edge shall be free from burrs and/or strands.

D. Visual cracks, chips, fibers and haloing from the inside edges of slots, notches, cutouts or holes shall not exceed 0.127 mm [0.005 inch] and shall be free from stray glass strands.

14. BOW OR TWIST

The bow or twist of the finished board shall be:

☐ Grade _____   (glass base material).

☐ Other _____

15. INSULATION RESISTANCE ☐ Required ☐ Not Required

Condition the complete board for 48 hours at $35.55° ± 1.11°C$ [$96° ± 2°F$] and $86 ± 5%$ relative humidity. The insulation resistance measured in this environment, with 500 volts DC applied between the two designated pads, shall equal or exceed ☐ 1000 megohms ☐ Other _____ megohms.

16. UL IDENTIFICATION ☐ Required ☐ Not Required

A. When a flame retardant base material is specified, the material shall have Underwriters' Laboratories (UL) recognition (minimum 94 V2).

B. The circuit board manufacturer's marking certifying UL recognition of the process (minimum 94 V2) is to be located as specified on sheet _____ of _____. The marking must consist of either the manufacturer's logo followed by full classification designation, such as 94 V1, or the manufacturer's logo followed by a type designation. When there is more than one factory, the factory designation must be included.

17. DIELECTRIC WITHSTANDING VOLTAGE ☐ Required ☐ Not Required

The dielectric withstanding voltage ☐ shall ☐ shall not be applied between conductor patterns of each layer and the electrically isolated pattern of each adjacent layer. The dielectric withstanding voltage shall be ☐ class _____ per IPC-ML-950 ☐ other _____

18. ELECTRICAL BARE-BOARD TESTING ☐ Required ☐ Not Required

All boards are to be 100% electrically tested for shorts and opens.

19. PACKAGING

The finished boards shall be ☐ individually ☐ bulk packed in unsealed thermo-plastic bags containing no silicones or sulfur-containing material which might degrade solderability.

20. APPROVED VENDORS

☐ Approved vendors per drawing 520830 Sheet-2 Vendor group ____

Figure 11.6 (*Continued*)

and orientation of all such permanently installed items. A drawing note should define the bonding and/or installation requirements. For example, a printed circuit board requiring a bonded heat sink must have the before-bonding cleanliness, location, orientation, and acceptable bow and twist after bonding defined on the master drawing.

It should be noted that the drawing may become too confusing if all this information is included on a single drawing or sheet. In such instances, a higher indentured drawing structure will be required or separate drawings should be prepared to show the part and the assembly operation.

11.12   Artwork

The styles of artwork are usually defined by their method of artwork generation. The three basic styles of artwork are style A1, documentation artwork; style A2, dimensionally stable artwork; and style A3, automated artwork. A1 style artwork is for reference only and all circuit topology is described in a nondimensional manner through reference to a grid or other technique to provide the relationship of conductors, lands, and other circuit configurations.

Style A2 is where the artwork is provided on a dimensionally stable master. The dimensionally stable master should be supplied or prepared on 0.008-in-thick stable polyester film, or equivalent, or photographic glass plates. The dimensionally stable master may be used in the development of the phototool.

All styles may be used in providing the conductor topology description interface between user and vendor. The accuracy of style A2 should be sufficient to permit direct reproduction of the production master and includes both reduction and step and repeat. Styles A1 and A3 artwork require additional processing other than photographic.

Intermixing of artwork styles in the documentation for a board, although discouraged, is permitted to the extent that consistency of the tooling provided to the manufacturer has the capability of producing the product within the tolerance limits defined on the documentation (master drawing, performance specification, etc.) package. Artwork may be generated for the following printed circuit board attributes:

- Circuit topology
- Solder mask
- Legend
- Solder paste stencil or screen

11.12.1   Documentation artwork

Documentation artwork (style A1) is used when the documentation facility wishes to have the manufacturer develop the artwork. In using this technique for description of the artwork, it is understood that the definition of artwork is for reference only and that all characteristics must be accurately dimensioned through note form or other techniques to ensure that the manufacturer's developed artwork meets the end-product requirements. All characteristics of grid location or other techniques to clearly define the circuit topology should be employed and clearly described on the artwork documentation master.

11.12.2   Dimensionally stable artwork

Style A2 for the development of artwork is one where the artwork is developed on a dimensionally stable master. The artwork may be at a 1:1 scale or may be at an enlarged scale, but should be capable of being used in the preparation of the phototool necessary to fabricate the printed circuit board. In some instances, the artwork may have its own drawing number; in other instances, the artwork may be part of a multiple-sheet master drawing.

The accuracy of the artwork master should be such that each has all centers of land areas, conductors, and other features located with respect to the true grid position within the positional tolerances established for that product. This dimensional requirement need not necessarily apply to the legend or permanent solder mask artwork masters.

11.12.3   Automated artwork (style A3)

All conductor definition and routing should be defined in the appropriate record format. Land patterns may be described explicitly, or by reference to a standard symbol library or symbol subroutine internal or external to the data job set. Comment records should be used to properly relate all critical features, artwork numbers, and revision level, and any other condition to the master drawing to which the A3 artwork data pertain.

11.12.4   Multiple-image artwork

In some instances, a multiple-image artwork may be delivered as the phototool as a part of the documentation package. This could be for any class of documentation. Since multiple-image artwork is normally tailored to a special manufacturing facility, process, or equipment, transfer of the manufacturing documentation to alternative manufacturers is limited. Quality conformance test circuitry should be included on the multiple-image artwork for each panel.

11.13   Manufacturing Documentation

Manufacturing documentation consists of information on tooling provided to supplement the engineering documentation. Manufacturing documentation should include, but not be limited to,

- Production masters
- Manufacturing layouts
- Soft tooling

11.13.1 Production masters

A production master supplied as a supplement to the engineering documentation should be a 1:1 master of a precision that will allow the product use or the production master in the manufacturing cycle. The production master may be a single- or multiple-image.

11.13.2 Manufacturing layout instructions

When a documentation package exists where complete manufacturing instructions are required, a production drawing should be developed which should be reviewed by the appropriate manufacturing facility for approval of all instructions to act as a manufacturing aid. In most instances, this document is used to provide the tooling locations of various manufacturing tooling marks and the step and repeat master information for registration of multiple-image artwork to the manufactured panel.

11.13.3 Soft tooling[2]

In light of the increasing sophistication of the printed circuit board manufacturer's equipment, it is now possible to transmit data in machine-readable form rather than the older, human-readable form. This is especially useful for data that will be used directly by another machine, such as a drill tape, and plotter tape.

Unless the printed circuit board design and documentation department has made prior arrangements with the printed circuit board manufacturer, the use of direct manufacturing tapes by the printed circuit board purchaser should be discouraged by the printed circuit board shop, unless it is a captive facility, because there may be some tooling built into the tape that will produce inefficiencies with the printed circuit board manufacturer. For example, a drill tape may contain pattern commands that are specific to one drill type and may not be compatible with the one to be used.

11.13.3.1 Drill tapes. Drill tapes will typically have one circuit and its coupon and will at least have the coordinates of all holes relative to the circuit datum. When drill tapes are provided with the documentation package, they should meet the requirements of IPC-NC-349. A drill tape must be verified. Thus, it is important to have the coordinates printed out and spot-check against the component-side artwork. If there is a mismatch, corrective action will have to be taken. Also, some tooling for insertion equipment may rely on these data. If so, this checking step becomes even more important.

11.13.3.2 Router tapes. The same considerations apply as for drill tapes. With routers, there is even less commonality of commands than with drills. In general, it may be more useful to generate the router tape from the board outline dimensions.

References

1. "Documentation Requirements for Printed Boards," ANSI/IPC-D-325, January 1987, Institute for Interconnecting and Packaging Electronic Circuits, Lincolnwood, Ill.
2. George R. Jacobs, Jr., American Electronic Laboratories, "Documenting Printed-Wiring Packages," *Machine Design*, May 15, 1969, pp. 166–173.
3. "Plastic Sheet, Laminated, Metal-Clad (for Printed Wiring Boards), Base Material GF, Glass Base, Woven, Majority Difunctional Epoxy Resin, Flame Resistant, Copper-Clad," MIL-P-13949/4, Revision C, February 1987.
4. "Standard for Foil Clad, Polymeric, Composite Laminate," IPC-L-112, July 1981, Institute for Interconnecting and Packaging Electronic Circuits, Lincolnwood, Ill.
5. "Flexible Metal-Clad Dielectrics for Use in Fabrication of Flexible Printed Wiring," ANSI/IPC-FC-241, Revision B, February 1986, Institute for Interconnecting and Packaging Electronic Circuits, Lincolnwood, Ill.
6. "Adhesive Coated Dielectric Films for use as Cover Sheets for Flexible Printed Wiring," ANSI/IPC-FC-232, Revision B, February 1986, Institute for Interconnecting and Packaging Electronic Circuits, Lincolnwood, Ill.
7. "Flexible Adhesive Bonding Films," IPC-FC-233, Revision A, February 1986, Institute for Interconnecting and Packaging Electronic Circuits, Lincolnwood, Ill.
8. "Specification for Rigid Substrates for Additive Process Printed Boards," ANSI/IPC-AM-361, January 1982, Institute for Interconnecting and Packaging Electronic Circuits, Lincolnwood, Ill.
9. "Metal Foil for Printed Wiring Applications," IPC-CF-150, Revision F (proposed), March 1989, Institute for Interconnecting and Packaging Electronic Circuits, Lincolnwood, Ill.
10. "Electroless Copper Film for Additive Printed Boards," IPC-AM-372, April 1978, Institute for Interconnecting and Packaging Electronic Circuits, Lincolnwood, Ill.
11. "Qualification and Performance of Permanent Coating (Solder Mask) for Printed Boards," IPC-SM-840, Revision B, May 1988, Institute for Interconnecting and Packaging Electronic Circuits, Lincolnwood, Ill.
12. "Printed Board Dimensions and Tolerances," ANSI/IPC-D-300, Revision G, January 1984, Institute for Interconnecting and Packaging Electronic Circuits, Lincolnwood, Ill.
13. "Automated Design Guidelines," ANSI/IPC-D-390, Revision A, February 1988, Institute for Interconnecting and Packaging Electronic Circuits, Lincolnwood, Ill.

About the Author

Gerald L. Ginsberg has written over twenty technical papers on printed circuit design and electronic equipment packaging. He is the author of *Surface Mount and Related Technologies*, and *A User's Guide to Selecting Electronic Components* and is a contributor to Coombs' *Printed Circuit Handbook*. Mr. Ginsberg has received Outstanding Achievement and Contribution awards from the Institute for Interconnecting and Packaging Electronic Circuits (IPC).

Index